D1063425
Urban Farming
2nd Edition
Sustainable City Living in Your Backyard, in Your Community, and in the World
Thomas J. Fox

CompanionHouse Books™ is an imprint of Fox Chapel Publishers International Ltd.

Project Team
Vice President–Content: Christopher Reggio
Editor: Amy Deputato
Copy editor: Jeremy Hauck
Design: Mary Ann Kahn
Index: Elizabeth Walker

ISBN 978-1-62008-301-7

Library of Congress Cataloging-in-Publication Data

Names: Fox, Thomas J., author.
Title: Urban farming : sustainable city living in your backyard, in your community, and in the world / by Thomas J. Fox.
Description: 2nd edition. | Mount Joy, PA : Fox Chapel Publishing, [2018] | Includes bibliographical references and index.
Identifiers: LCCN 2018032077 (print) | LCCN 2018037304 (ebook) | ISBN 9781620083024 () | ISBN 9781620083017 (softcover)
Subjects: LCSH: Urban agriculture. | Sustainable agriculture. | Vegetable gardening.
Classification: LCC S494.5.U72 (ebook) | LCC S494.5.U72 F69 2018 (print) | DDC 635.9/77--dc23
LC record available at https://lccn.loc.gov/2018032077

Fox Chapel Publishing
903 Square Street
Mount Joy, PA 17552

Fox Chapel Publishers International Ltd.
7 Danefield Road, Selsey (Chichester)
West Sussex PO20 9DA, U.K.

We are always looking for talented authors. To submit an idea, please send a brief inquiry to acquisitions@foxchapelpublishing.com.

Printed and bound in China
21 20 19 18 2 4 6 8 10 9 7 5 3 1

Contents

Dedication

To my mom, for a love of writing; my dad, for a love of gardening; my wife, for a love of life; and my boys, with great love and hope for a greener future.

Acknowledgments

Thank you to Andrew DePrisco, who opened the door to the book; to Karen Julian, who ushered it through; and to my two crack editors, Jennifer Calvert and Amy Deputato, who lifted the peaks and filled in the valleys with extraordinary skill and restraint. Derek Burnett deserves tremendous credit for his expert help on this book, as well as other projects. I'd also like to acknowledge the many people who tolerated my ignorance with grace, taught me so much, and shared their enthusiasm for urban agriculture, among them Martin Bailkey, Rick Bayless, Don Boekelheide, Natalie Brickajlik, Fred Brown, John Cannizzo, Roxanne Christensen, Virginia Clarke, Mary Seton Corboy, Daniel Dermitzel, Wes Duren, Danielle Flood, Lorraine Gibbons, Carole Gordon, Jennie Grant, Mike Hamm, Sherilin Heise, Gregory Horner, Jonathan Jones, Jerry Kaufman, Aley Kent, Erik Knutzen, Michael Levenston, Andrew McCaughan, Michael McConkey, Joe Nasr, Molly Philbin, Robert Philbin, Gordon Prain, Jessica Prentice, Martin Price, Brooke Salvaggio, Wally Satzewich, Bob Scallan, Mike Score, Jill Slater, Jac Smit, Lena Carmen Soileau, and Brenda Tate. It would have been a much poorer book without them, and a much less enjoyable task of writing.

INTRODUCTION

In January 2009, I answered an ad posted with the Editorial Freelancers Association seeking a writer for a book on city living/urban farming. I jumped at the chance. At the time, I'd spent most of my adulthood living in cities, and I'd written about and for environmental organizations based in them. Some of my most vivid memories of my childhood were the cucumbers and asparagus my father grew in the narrow stretch between our suburban homes.

The person who had posted the ad wrote back to ask what I'd write about. I pitched an outline and was eventually chosen to write the book.

Still...*urban farming*. Back then, "urban farming" was like "salad couch." Sure, you could smoosh the two words together, but what the hell would you be talking about?

Of course, just because the idea of urban farming might have struck Americans as odd in the early twenty-first century, that's not to say that the practice of urban farming was new. It was woven throughout the history of North America, most famously in the Aztec capital of Tenochtitlan, now known as Mexico City. It was practiced throughout world history, in fact, and wasn't just a thing of the past. Greensgrow Farm has thrived in Philadelphia since 1997, the Dervaes family has maintained an urban homestead in Pasadena since 1985, and City Farmer has been doing it in Vancouver since 1978. And those are just three well-known examples among countless others.

Indeed, I discovered that urban agriculture had been seriously considered by some in the development community since 1996, when Jac Smit and Joe Nasr published *Urban Agriculture: Food, Jobs and Sustainable Cities* for the United Nations Development Program (UNDP). But somehow the terms "urban farming" and "urban agriculture" hadn't quite made it to American public consciousness—

Talk about local color! Beautiful local produce adorns farmers' markets across the nation and around the world.

at least based on the blank stares I received whenever I would tell people I was writing a book on urban farming.

The concept seemed at best exotic, if not entirely oxymoronic. There didn't seem to be much in the way of nonacademic books squarely addressing the topic. Growing food was just one aspect of Kelly Coyne's and Erik Knutzen's *Urban Homesteading*, which covered a wider set of lifestyle issues, while other existing books applicable to city farmers focused more narrowly on some combination of technical or philosophical issues, like Mel Bartholomew's *Square Foot Gardening* or H.C. Flores's *Food Not Lawns*.

In other words, I hoped to make a minor contribution to a very obscure field about which almost nothing seemed written outside of international development organizations and other specialists. A few months later, while I was still researching this book, Novella Carpenter's *Farm City: The Education of an Urban Farmer* came out ("Fresh, fearless, and jagged around the edges," as Dwight Garner described it in the *New York Times*). The month before my manuscript was due, Kelly Coyne and Erik Knutzen published a revised and expanded edition of *The Urban Homestead* (I profile Kelly and Erik in Chapter 5). Dickson Despommier's *The Vertical Farm* arrived in between my manuscript submission and its publication, which was itself quickly followed by Annette Cottrell and Joshua McNichols's *Urban Farm Handbook*, Jennifer Cockrall-King's *Food and the City: Urban Agriculture and the New Food Revolution*, Sarah C. Rich's *Urban Farms*, and an ongoing stream of additional books by farmers, journalists, foodies, and others. There's probably been one published since you started reading this.

Still, the urban farming bookshelf was and remains roomy and diverse enough to embrace a book that reflected my interests in both the larger social issues that led to this historical moment for urban farming, especially in North America, and how one could engage in it. In other words, a book that explores two questions about urban farming: "why now?" (Part I, Chapters 1–3) and "how?" (Part II, Chapters 4–9).

I'd like to say that my book helped us usher in a revolution, but the fact is that I was lucky enough to crest a wave. And what a wave it has been. As Michael Levenston of City Farmer in Vancouver, Canada, notes, urban agriculture has gone from back page news to front page news. It's no longer salad couch. Levenston points

A street market in Bali may look different from the farmers' market down the street, but they're both built on the same principles.

to an urban farm being a central plot device in the web comedy series The North Pole. Several podcasts center around urban agriculture and farmers. Consumer giant Unilever even launched an entire "Growing Roots" vegan snack line that plows half of its profits into urban farming.

Changes have happened so quickly that it's been a challenge deciding what to update in this edition, and how. In the end, the lion's share of updates fall within Part I, the "why now?" section of the book.

Chapter 1 (Feeding Our Cities) largely holds up as is, though we continue to experience events related to national security. For example, Hurricanes Harvey and Irma underscored for many American how climate change has exacerbated the risk of such storms and made resilient food systems more important. As of this writing, Cape Town, South Africa, is poised to become the first major city in modern times to run out of water; I doubt it will be the last. I did not reiterate how urban agriculture, as part of a more sustainable food system, advances national security because, unfortunately, we will continue to be struck by new examples of this again and again.

Likewise, the fundamentals of many of the examples discussed in Chapter 2 (Marching to Sustainability on Our Stomachs) are relatively evergreen.

Did You Know?

In one of the most telling sign of urban agriculture's arrival, it featured for the first time in the USDA's 92nd Agricultural Outlook Forum in February 2016.

I made the most changes in Chapter 3 (Toward an Urban Farming Future). While many of the biggest challenges for urban farmers remain the same—land costs, transactional costs, and issues with zoning and other municipal policies—tremendous headway has been made. Thanks to pioneering urban farmers and farsighted policymakers, for example, there has been a wave of newly pro-urban agriculture policies and reforms, including in Minneapolis and San Diego (2012); Boston and St. Paul (2013); Atlanta, Aurora (Colorado), San Francisco, and Spokane (2014); Pittsburgh, Sacramento, and Savannah (2015); Flagstaff and Indianapolis (2016); and Fargo, Laredo (Texas), and Los Angeles (2017).

Urban agriculture has gained wider expert acceptance. Take certificate and degree programs in urban agriculture, for instance, which are currently offered by the University of Illinois, University of San Francisco, Kansas State University, City Colleges of Chicago, San Diego City College, Prince George's Community College, Ryerson University, Virginia State University, Purdue University, the University of Colorado, the University of the District of Columbia, and the University of Florida—and probably even more schools by the time this book is published. And that's not including the scads of institutions offering certificates and degrees in food systems, sustainable agriculture, and related topics.

Urban farming has become, if not ubiquitous, then at least respectable. It is a legitimate topic of academic inquiry. For example, a 2016 study by Carolyn Dimitri, Lydia Oberholtzer, and Andy Pressman examines the social missions of urban farmers in the United States based on a survey of hundreds of them. Another study, spearheaded by researchers at Arizona State University and Google, Inc., and published in 2018, uses huge datasets to estimate the actual and potential value of urban agriculture worldwide. While I was not able to benefit from this research when writing this book, I have referenced it in this updated edition.

Several large commercial ventures have flourished. AeroFarms operates several aeroponic farms, including the world's largest indoor vertical farm, in New Jersey. Brooklyn-born Gotham Greens has steadily been expanding its trademark rooftop greenhouses, with three in New York (at 15,000, 20,000, and 60,000 square feet) and the latest in Chicago (75,000 square feet, making it the world's largest rooftop farm as of this writing). Philadelphia's Metropolis Farms has been perfecting technology for hydroponic vertical farms that can grow anything from cannabis to strawberries while, 250 miles to the west, Pittsburgh provides a home to Hilltop Urban Farm, currently billed as the nation's largest urban farm.

Part II—the "how to?" section—remains largely the same in this updated edition. Choosing which plants to grow and how best to do so borrow on traditions going back millennia. The biggest changes to Part II concern technological improvements

in terms of containers (Chapter 4: Starting Your Farm) and lighting (Chapter 8: Plant Management).

The urban farms and farmers I profiled throughout the original edition posed a challenge for me with this updated edition. The landscape of urban agriculture is fluid, with new farms opening, closing, or evolving all the time. For example, BADSEED Farm, profiled in Chapter 6, transformed, chrysalis-like, into to the even larger Urbavore at a new location in Kansas City, Missouri. The controversial Hantz Farms—which had aspired to be the world's largest urban farm (see Chapter 3)—has pivoted into a huge (and still controversial) urban agroforestry project known as Hantz Woodlands. And so on.

Ultimately I chose to keep the original profiles in this edition because the urban farmers' experiences at the point in time captured in the book could be those of their peers elsewhere in North America today. I added a new afterword that includes a profile of Hilltop Urban Farm and an updated profile of Brooke Salvaggio and Daniel Heryer so readers can trace nearly a decade of their evolution as urban farmers from BADSEED to Urbavore.

Finally, I've revised the recommended resources, and I now host a continually updated list on my website, www.thomasjfox.com.

Part I

The Big Picture

CHILLI
COCONUT
SAFFRON
CURRY
PEPPER

CHAPTER 1

Feeding Our Cities

In towns and cities across the globe, in large ways and small, urban farming is quietly gaining momentum. If you're slurping a bowl of hot tom yam goong from a street vendor in Bangkok, enjoying a traditional potato omelet (chips mayai) in Dar-es-Salaam, sipping a glass of merlot in Santiago, or indulging in honey-and-goat-cheese ice cream at the Fairmont Waterfront Hotel in Vancouver, chances are you are supporting urban farming. Modern urban farming is closely connected with urbanization, and increasingly with a conscious move toward sustainability. It has even become an unexpected necessity in some places, such as Havana (pictured).

The human population of the world is rising by about 75 million people per year—mostly in cities—and is expected to exceed 9 billion by 2050. Sure enough, some of the growth in urban farming happens when towns grow into cities, and cities into megacities, sprawling into once-rural land. Instead of displaced rural farmers working the newly urban landscape, researchers have found that most urban farms are established by city dwellers. It is usually driven in the global north by those looking to reconnect with a sense of place and to live more sustainably, and in the global south by those just looking to live.

Across the United States, communities are taking steps to create a more welcoming atmosphere for agriculture through farmers' markets, zoning-law changes, and use of underused green spaces and brownfields (former industrial sites), often through the irrepressible efforts of a few individuals with a passion to make it happen. One such example is the Goat Justice League in Seattle, which fought to legalize goats within the city limits and succeeded with pygmy goats. But is farming in the city even realistic? The short answer is *yes*.

According to a data-driven assessment of urban agriculture by Arizona State University and Google published in 2018, an estimated 5 to 10 percent of the world's noncereal crop production is already produced in and around cities. It found that the "[r]ecommended consumption of vegetables for the urban population may be met almost entirely" through urban agriculture and did not even consider urban livestock. Many individual countries and cities are even more advanced. Shanghai (pictured), for example, produces more than 50 percent of its consumed chicken and pork, 90 percent of its eggs, all of its milk, and more than 2 million tons of wheat and rice in and around the city. And Shanghai is no shrinking-violet, backwater city—it has roughly 20 million residents and more than four times as many skyscrapers as Manhattan.

Even as urban agriculture has taken root in cities around the world, traditional rural agriculture—at least the Currier & Ives vision of it—has evolved into something more Dickensian. The changes in farming over the past three centuries have brought extraordinary productivity, both enabling and enabled by growing cities. However, only recently has the true cost of these gains emerged. At its worst, this "industrial agriculture" is antithetical to our heritage, as discussed in the next section, and a threat to our future.

Roots of Urban Farming

In March 2009, in the midst of a recession and two wars, First Lady Michelle Obama helped break ground on a new vegetable garden at the White House—the first since her predecessor Eleanor Roosevelt planted a "victory garden" in the midst of

First Lady Michelle Obama in the White House vegetable garden.

World War II. Mrs. Roosevelt's garden had itself hearkened back to the work-relief gardens of the Great Depression. Before that came the Federal War Garden program of World War I as well as Detroit's "potato patches" and other responses to the 1893 depression. Urban dwellers have turned to gardens countless times throughout history as a way of weathering adversity and regaining a sense of autonomy.

The White House is not alone. Since 2009, statehouses and municipal governments from Baltimore to Sacramento have begun their own food gardens. The United States Department of Agriculture opened a "People's Garden" at its headquarters and encouraged similar efforts at its facilities around the country. Seed sales jumped by about 25 percent, and about 40 percent more households grew vegetables that year than two years earlier.

The same holds true cross the Atlantic. In June 2009, Queen Elizabeth unveiled a vegetable patch on the grounds of Buckingham Palace, the first (once again) since World War II. The waiting list in London for allotments—patches of land rented out to gardening-minded residents at a nominal cost—can stretch into decades, and the supply of allotments in the United Kingdom is reportedly short about 200,000 units, in a country with one-fifth the US population. Nevertheless, inspired by Vancouver's success establishing 2,010 new garden plots in time for the 2010 Olympics, London created 2,012 new agricultural spaces in time for its own Olympics two years later. New construction throughout the European Union may soon include integrated "vertical allotments" in accord with regulations being considered by the European

PEOPLE'S GARDEN CONCEPT PLAN

PHASE ONE ORGANIC VEGETABLE GARDEN

USDA WHITTEN BUILDING WASHINGTON DC

MARCH 2009

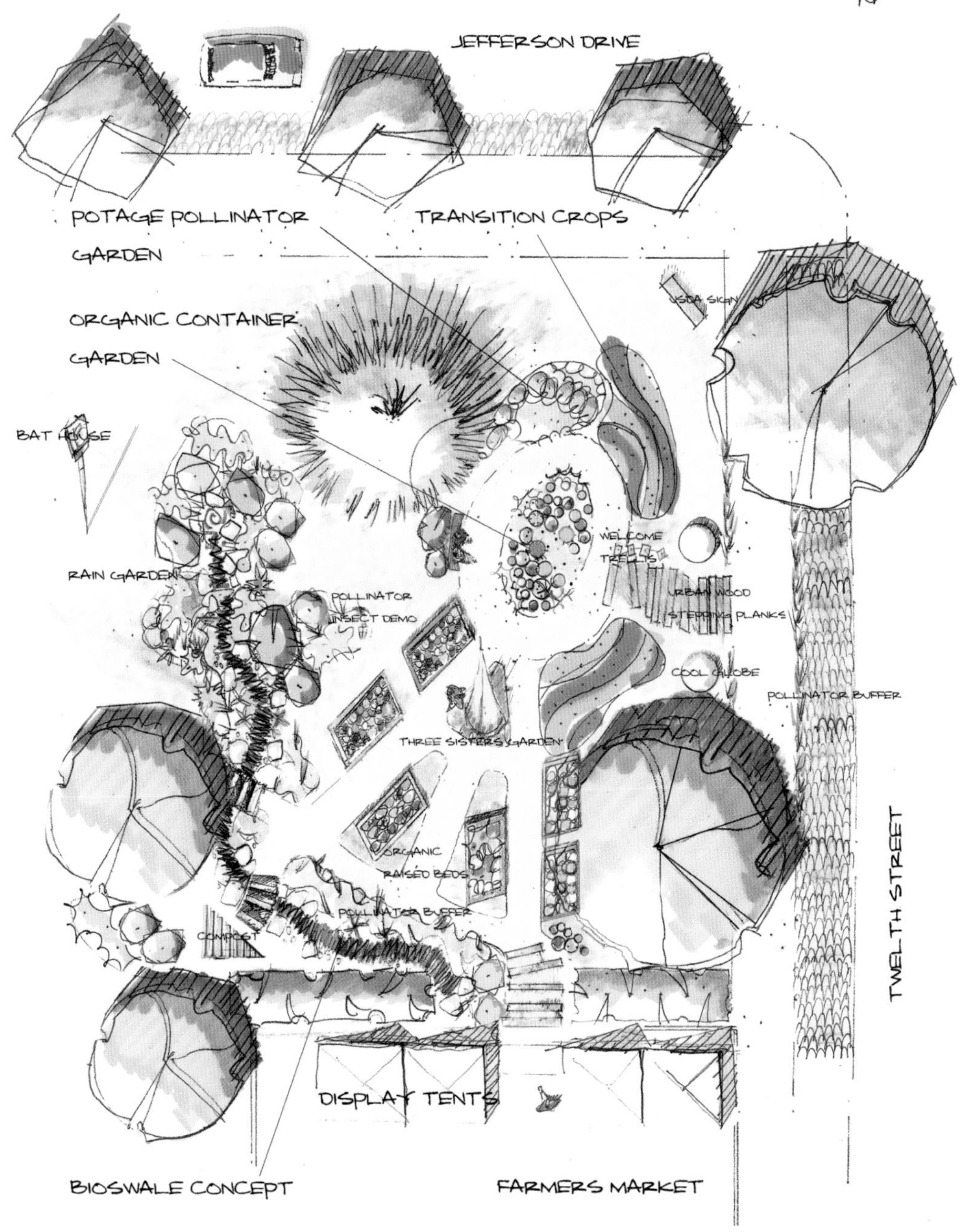

The perennial fascination with the Hanging Gardens of Babylon speaks to the enduring allure of urban agriculture.

Environment Agency. These allotments could include balconies, rooftops, and walls earmarked for growing food on high-rise buildings.

Why is urban farming integrated into cities such as Shanghai but still a novelty in the United States? Certainly, part of the reason is that we have profited so abundantly from the transformation from traditional farming into industrial agriculture—yields per farmer have skyrocketed. This success has reinforced the notion that city is city and country is country, and never the twain shall meet—except in supermarket aisles. It is a bias evidenced, perhaps, by the fact that goats in Seattle may be more striking to us than a world population ballooning beyond the ability of conventional agriculture to feed it. Yet this separation of *urban* and *farming* is a modern one.

The histories of cities and agriculture are, in fact, inextricably linked. Historians may bicker about whether the discovery of agriculture encouraged our ancestors to settle down into permanent settlements, or whether the first settlers developed agriculture out of necessity, but the correlation between the two is clear. Some of the plants and animals first domesticated were cultivated in the rich soil of ancient

Fertile Crescent cities such as Jericho (West Bank), Damascus (Syria), Susa (Iran), Tyre (Lebanon), and Catal Huyuk (Turkey).

Egypt and the city-states of Mesopotamia had developed advanced, irrigated agricultural techniques by 6,000 BC, some possibly employed in the legendary Hanging Gardens of Babylon (which might have actually been in Nineveh or Nimrud; all three cities are in modern-day Iraq). These Near Eastern civilizations also dabbled in aquaculture—the farming of seafood—as did ancient China, which continues the practice on a large scale. In fact, China has long practiced advanced agricultural techniques to feed its many towns and cities, maintaining a stronger connection to urban farming than most places in the world. Then, as now, China also employed an "aqua-terra" system of wetland farming, which was familiar to ancient Indonesia as well.

One of the most famous historical examples of urban farming occurred in and around Tenochtitlan, the Aztec capital that is now Mexico City. On shallow lake bottoms, the Aztecs built *chinampas*, which were essentially raised beds fenced in with woven canes—almost giant baskets—filled in with river mud and organic matter to above the water level. Aztec farmers traveled between rows of chinampas by boat. Hundreds of miles south and thousands of feet higher, the Incas built farming terraces into mountains and their cities, such as Machu Picchu. The Mayans practiced urban agriculture extensively as well.

When Rome finally defeated its nemesis city-state Carthage (in modern Tunisia) in 146 BC, legend has it that the Romans plowed salt (a plant killer) into the ground so that nothing would grow there, essentially erasing a city by destroying its agricultural base (well, by that and by killing or enslaving the entire population and reducing the city to ashes). It's improbable that the Romans would actually sprinkle the fields with salt—it was an expensive commodity—but the legend points to an appreciation of urban farming.

Unfortunately for Rome, this appreciation didn't translate into action. In a funny twist of fate, the Roman politician keenest to have Carthage destroyed, Marcus Porcius Cato, also had a major beef with lazy Romans who "prefer to exercise their hands in the theatre and the circus rather than in the corn field and the vineyard." They may have followed his advice in destroying Carthage, but they completely ignored his enthusiasm for farming. As a result, Rome's population growth, soil depletion, and reliance on imported food contributed to its own downfall six centuries later.

The rises and falls of great cities—and civilizations—have long been intimately tied to agriculture. (And as Rome discovered, sometimes karma really is a boomerang.)

So What Happened to Urban Agriculture?

The long decline of urban agriculture coincided with technological advances of the Industrial Revolution (or Revolutions, according to some), which brought with them a change in perception about the roles of cities, rural communities, and agriculture. A passage from historian Will Durant is revealing, written roughly halfway between the post-Civil War flowering of the Industrial Revolution in the United States and today:

> **The first form of culture is agriculture. It is when man settles down to till the soil and lay up provisions for the uncertain future that he finds time and reason to be civilized. Within that little circle of security—a reliable supply of water and food—he builds his huts, his temples, and his schools; he invents productive tools, and domesticates the dog, the ass, the pig, at last himself. He learns to work with regularity and order, maintains a longer tenure of life, and transmits more completely than before the mental and moral heritage of his race....Culture suggests agriculture, but civilization suggests the city.**

Durant manages two great insights here: one intentional and one not. His understanding of the connection between agriculture and cities reflects an ancient sensibility, yet he also reveals a modern industrial bias that views cities as the zenith of civilization—one that excludes agriculture, or at least pushes it to the rural fringes. It is a tendency to view the city as something that has transcended agriculture, a concept that civilization may have sprouted in the field but only blooms in the boardroom.

This attitude pervades our culture, even in the very sciences responsible for feeding people. Noting that very little agricultural research concerns urban agriculture, Gordon Prain, the global coordinator of Urban Harvest (an initiative of the Consultative Group on International Agricultural Research [CGIAR]), writes that the disparity "is related to the sectoral separation of 'urban' and 'rural,' a separation that has its roots in the Industrial Revolution and its subsequent transfer through colonial expansion to the developing world."

The Industrial Revolution was punctuated by a flurry of world-changing inventions and discoveries within a relatively short period, among them the steam engine, the Bessemer process for making steel, the rediscovery of concrete, pasteurization, and all kinds of machines. In the United Kingdom, its epicenter, the Industrial Revolution went hand-in-hand with a series of "Inclosure Acts." These laws divided up "the commons" (lands everyone could share for farming, pasturing livestock, gathering wood, and other purposes). New farming methods required large fields, which the authorities created by consolidating the commons' traditional crazy quilt of small plots and fencing them off. By eliminating traditional rights to share the land, the laws had the effect of taking away the livelihood of rural peasants ("commoners"), who comprised the majority of farmers.

No longer able to survive in the country, displaced peasants flooded into cities to feed the new craze: making stuff. As a result, rural food production and urban manufacturing in mills and factories both exploded. The mass production of goods—edible and otherwise—of the new era coincided with mass consumption made possible by colonial expansion, booming population growth, and improvements in transportation. The Industrial Revolution also oiled the economic machine, providing a world stage for corporations and trade unions, which entered stage right and stage left, respectively.

Industrialization prompted the divorce of urban and rural, with rural getting sole custody of agriculture—an arrangement that remains the status quo. As discussed next, however, this rigid distinction has outlived its usefulness, and the results are ever more disastrous.

The New Business of Agriculture

The tectonic agricultural shifts of the Industrial Revolution reached earthquake intensity in the latter half of the twentieth century. In particular, a "Green Revolution" began after World War II, prompted by peace and a desire to feed a growing world, and enabled by new high-yielding crop varieties, irrigation techniques, and synthetic pesticides and fertilizers—starting with a postwar American surplus of ammonium nitrate, an ingredient in explosives. Cheap oil and water fueled the revolution. In the United States, the practice of farming evolved into "agribusiness" thanks to economies of scale, government subsidies, and an official bias best captured by the mandate of Earl Butz, Secretary of Agriculture under Presidents Nixon and Ford: "Get big or get out."

And so farmers did. Mary Hendrickson and William Heffernan of the University of Missouri have tracked this consolidation in terms of the *concentration ratio*, or how much of the total market the top firms in each industry control. For example, a 2007 study showed that the top four players in beef packing controlled 83.5 percent of the market, and the top four companies in pork packing controlled an estimated 66 percent. In flour milling, the top three companies controlled 55 percent of the market. And we're not talking about eleven different firms. We're talking about seven, because some companies dominate in more than one industry. Cargill, for example, is a leader in all three categories.

At its worst, this concentration in agribusiness has resulted in industrial agriculture, of which the Union of Concerned Scientists (UCS) outlines four main characteristics: monoculture, few crop varieties, reliance on chemical and other inputs, and separation of animal and plant agriculture.

Monoculture at its prettiest: fields of corn as far as the eye can see.

Monoculture

Monoculture is the cultivation of a single kind of crop in a given area. Our current agricultural system has immense swaths of monoculture, including our "amber waves of grain." Among the principal crops tracked by the National Agricultural Statistics Service, for example—mainly grains, legumes, sugar crops (cane and beet), and tobacco—just three made up 70 percent of the US acres planted in 2009: corn, soybeans, and wheat.

Monoculture contrasts with polyculture, the multiple-plant system that typically characterized preindustrial farming traditions worldwide. Many Native Americans, for example, planted the "three sisters" (corn, beans, and squash) together. There are many advantages to polyculture, but one of the most fundamental is that it spreads out the risk of crop loss—from weather, disease, weed competition, or animal pests—among plants with different susceptibilities. An outbreak of southern corn leaf blight, for example, might destroy a season's worth of corn, but leave the bean and squash harvests intact. As the UCS points out, that very disease destroyed 60 percent of the US corn crop in 1970. Similarly, China's fondness for poplars' strengths—fast growth and easy propagation—led to hundreds of miles of trees also sharing the same weaknesses, such as the Asian Longhorn Beetle. In 2000, this single variety of pest dealt a catastrophic blow to China's reforestation efforts by reportedly killing 1 billion poplars.

Monoculture is the agricultural equivalent of investing in a single stock. You do well in favorable times, but in bad times, your portfolio can totally tank. It's a dangerous strategy when you depend on your portfolio to eat.

Limited Varieties

Our monocultural practices are particularly dangerous in that they don't just involve a limited number of crops—corn, wheat, and the like—but a limited number of varieties within those species, magnifying the risk of catastrophic losses. Most food crops that have sustained people for millennia have several—even hundreds—of cultivated varieties, or *cultivars*. They are the same species but are distinguished by a fairly uniform collection of varietal traits. You can think of Chihuahuas and Rottweilers as different cultivars of dog: they're both the same species, but they have been bred to exhibit very different characteristics, including strengths and vulnerabilities.

The almighty Russet Burbank potato: an example of dominance built on shaky feet.

In the United States today, however, about 90 percent of the soybeans and two-thirds of the corn, constituting just a handful of varieties, is genetically engineered to resist pests or herbicides. Out of thousands of potato varieties, just one—the Russet Burbank, preferred by McDonald's for its French fries—dominates more than half of the world's potato crop, according to the UCS. In addition, most livestock now comes from a limited number of gene lines, since predictable uniformity makes growth rates, feed requirements, and automated processing more efficient.

Industrial agriculture's focus on specific varieties hits heirloom varieties of plants and animals particularly hard. These are older varieties that reproduce true to type, such as the Maori Kunekune pig, the Stayman Winesap apple, or the butterscotch calypso bean. Such special animals and plants once would have been cherished and passed from generation to generation of farmers. As food production grows ever more consolidated and focuses on a narrow array of crop varieties, many of these heirlooms simply disappear.

Chemical Inputs

According to the Fertilizer Institute, the United States is a "mature market" for fertilizer, meaning that annual demand remains fairly steady. According to the Institute's website, in 2004, the United States used about 57.8 million tons of fertilizer, of which 23.4 million tons were the major nutrients nitrogen, phosphorous, and potassium (potash).

We have also generally plateaued in the use of pesticides—which includes herbicides, insecticides, and fungicides—at about 1 billion tons per year. Agriculture comprises about three-quarters of the use of pesticides, mostly herbicides. (One reason herbicides may be dominant is that some crops are genetically engineered to be herbicide resistant so that a whole field can be sprayed, killing weeds but not crop plants.)

Livestock that are "factory farmed" often present a triple-whammy of inputs: the fertilizer used to increase production of feed crops, an array of pesticides used to maximize yield, and antibiotics to promote growth and prevent infection. In fact, the UCS estimates that agricultural uses account for about 70 percent of the

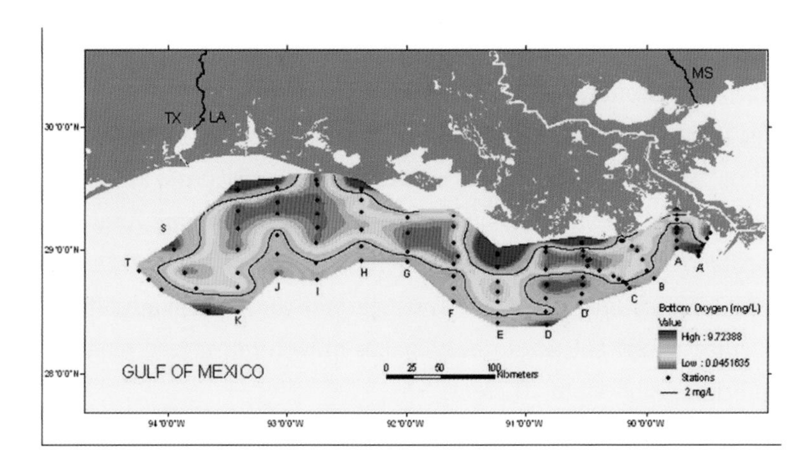

A giant dead zone in the Gulf of Mexico, attributed to agricultural runoff traveling down the Mississippi River.

antibiotics used in the United States. Cows might also receive growth hormones to boost milk or beef production.

Pesticide runoff is implicated in killing or mutating wildlife, while nitrogen- and phosphorous-rich fertilizer runoff damages aquatic ecosystems in a process called *eutrophication*. The extra nutrients cause an explosion in algae populations—often enough to turn the water red or brown—followed by a huge die-off when the algae have eaten all of the nutrients. Decomposition of the dead algae sucks up dissolved oxygen, resulting in a hypoxic environment that kills fish, mollusks, crustaceans, and most everything else we consider seafood. The largest dead zone off the United States (at times as large as the state of New Jersey) is in the Gulf of Mexico, into which the Mississippi River flows after coursing through "America's breadbasket."

In addition, agricultural use of antibiotics may spur resistance among bacteria, which is bad news for us. Hormones given to cattle may likewise pose health risks to humans and other animals; the use of hormones is banned in the European Union.

Animal/Plant Separation

One of the most unusual characteristics of industrial agriculture, at least from a historical perspective, is the separation of plant and animal agriculture. We used to have what felt like the perfect arrangement: livestock such as cows and goats would mow the lawn, eat the harvest leftovers, and graciously provide milk and fertilizer in exchange. Chickens would eat bugs and weed seeds in exchange for eggs and more fertilizer. Elton John would sing about the circle of life. More common these days, however, is for field crops and livestock to be separately concentrated, maybe even far from each other.

According to the UCS, a mere 5 percent of farming operations now generate 50 percent of the livestock produced in the United States. We're not talking about a large herd of cattle lowing through the valley or a feathery cloud of free-range chickens. That's not a sufficient scale. No, concentrated animal feeding operations (CAFOs is the technical term) are big. Typical CAFOs might have 10,000 cattle, 25,000 pigs, or 100,000 chickens. They are, in essence, livestock cities thrown up in rural settings, and with some of the worst problems that affect human cities—particularly waste. In 2005, the dry weight of livestock waste exceeded 335 million tons—more than a ton for every person in the United States.

So What?

Though industrial agriculture has many philosophical detractors—such as locavores (people who eat locally produced food), animal rights groups, workers' rights groups, and people downwind of CAFOs—perhaps the most common objection is that it is simply unsustainable, both economically and ecologically. One might fairly counterargue that never has agriculture produced so much food so cheaply for so many. Between 1920 and 1999, for example, US corn yields per acre increased by nearly 350 percent.

Production increases alone do not signal sustainability, however, especially when that productivity depends on finite resources. According to the National Sustainable Agriculture Information Service (formerly the Appropriate Technology Transfer for Rural Areas project, but still called ATTRA):

> **The industrial approach, coupled with substantial government subsidies, made food abundant and cheap in the United States. But farms are biological systems, not mechanical ones, and they exist in a social context in ways that manufacturing plants do not. Through its emphasis on high production, the industrial model has degraded soil and water, reduced the biodiversity that is a key element to food security, increased our dependence on imported oil, and driven more and more acres into the hands of fewer and fewer "farmers," crippling rural communities.**

Even without climate change, contemporary agriculture's dependence upon oil, cheap water, and synthetic inputs would be unsustainable. Yet we do also have climate change to deal with, and it does not bode well for most agriculture in most places. Increased heat can stress many commonly grown crops. Every 1-degree-Celsius (almost 1.8-degree-Fahrenheit) increase in temperature, for example, reduces yields of wheat, rice, and corn by about 10 percent. Increasing atmospheric carbon dioxide levels reduce the efficacy of one of the most important herbicides, while simultaneously giving a boost to weed growth—particularly that of poison ivy—according to researchers at Duke University. The ivy grows faster, gets bigger, and becomes more toxic at higher levels of carbon dioxide.

In the future, you may be seeing a lot more of this: poison ivy.

Whole civilizations have collapsed when their once-successful agricultural schemes failed them. Sometimes this has happened relatively suddenly. Easter Islanders so deforested their island within several centuries that it was virtually treeless by the time Europeans first visited in the early eighteenth century. Often it has happened gradually. Mesopotamian and Classical Mayan civilizations both declined after they slowly destroyed the fertility of their soils through poor management of irrigation—a practice key to the Green Revolution and US agriculture.

Our own civilization has suffered the collapse of commercial fisheries, such as Pacific sardines in the 1970s, Atlantic cod in the early 1990s, Pacific groundfish in the late 1990s, and Pacific salmon in the early twenty-first century. We are fortunate to have so many natural riches to survive the losses—arable land, plentiful lakes and rivers, and several fisheries—but the losses are still significant. The combined catch of their (overharvested) peaks total about 1.5 billion pounds. These collapses cost not only a lot of high-quality seafood but also tens of thousands of jobs and four key holdings in our portfolio of food security.

As destructive as overharvesting can be—whether trees, fish, or otherwise—the human activities likely to cause the greatest negative effect on foodstocks between now and 2100 are those that contribute to global climate change. It goes far beyond the threats facing particular fisheries, breadbaskets, or other food sources.

Agriculture plays a major role in the release of greenhouse gases. Livestock directly contributes somewhat—as through cows belching—but the real culprit is fossil fuels: as raw ingredients in agrochemicals; in the manufacture of agriculture-related plastics; in gasoline for tractors and transport trucks, jet fuel for aircraft, and bunker oil for container ships; in the refrigeration of perishable items; and in countless other uses. As a result, agriculture may only produce 15 percent of the world's carbon dioxide emissions—and we need to eat, after all— but it also produces about half of the human-sourced methane and two-thirds of the nitrous oxide, both of which are far more potent than carbon dioxide.

Irrespective of its capacity to feed the world, the way the United States (and much of the developed world) "does" food is simply unsustainable on many levels. And, as we'll see next, its capacity to feed a growing world by itself is doubtful.

The Renaissance of Urban Farming

Urban agriculture has probably never gone out of style in places where food was historically scarce or unreliable, such as China. It may be novel in places where urbanization itself is new, as in parts of Africa. In much of the rest of the developed world, however—including the Americas, Europe, and large parts of Asia—farming was something many felt that cities had moved beyond. For us, what urban farming now enjoys is a renaissance. Whatever one's local view of urban farming, three broad currents coalesce in answering "why now?"—the constraints of the present, a fear of the future, and a nostalgia for the past.

A Shrinking Planet

To feed a growing population, global food production will need to increase by roughly 50 percent by 2030 and double by 2050. Historically, people have met an increased demand for food by either increasing the amount of land under cultivation, increasing productivity per unit of land, or both.

A relatively optimistic joint report from the Organisation for Economic Co-operation and Development (OECD) and the Food and Agriculture Organization (FAO) of the United Nations finds that the world is currently using only about 50 percent of the total arable land not directly competing with other important uses (such as cities or forests). So there may be more land, but most of the unused, rain-fed arable land is in sub-Saharan Africa or South America—only a partial match for where most of the population growth will be. That means we'll need vast amounts of fuel to bring the food to market. Even assuming that transportation costs pose no problem (which, of course, they do), there's a limit as to how quickly unused arable land can be converted to crops. Between 1960 and 2010, the amount of arable land increased by less than 11 percent, and—thanks to increased cropping intensity—the harvested area expanded by about 24 percent, but the total world population more than doubled.

The sustainability of currently farmed land poses other challenges. Agricultural tillage—the act of physically loosening or breaking up the soil, as with a plough, harrow, or hoe—contributes greatly to the problem. It is traditionally done to improve aeration, control weeds, and release a burst of nutrients from the soil. The problem, unfortunately, is that it greatly contributes to soil erosion. Loose soil blows more easily in the wind and flows more readily with the rain. In drier climates, poor farm management or overgrazing can lead to desertification, which is essentially turning once-arable land into desert.

You might not think that wind and water would sweep away so much dirt, but United States croplands annually lose, on average, over 4 tons of topsoil per acre. In 2003, for example, this translated to about 1.5 billion tons—almost 12,000 pounds (5,443 kg) for every person in the nation. Since it can take 1,000 years to form an inch of soil—and topsoil may be just 6 inches thick—allowing this level of erosion clearly makes food production unsustainable.

Increasing productivity is a challenge, as well. Strong efforts in the developing world can create productivity gains, but the productivity of croplands in developed countries is only incrementally increasing, hitting a plateau, or—according to some—possibly even declining. Lester R. Brown of the Earth Policy Institute noted that in a recent span of nine years, world grain production fell short in six of those years, and world stocks of grain at the outset of the 2008 harvest were at near-record lows. Brown pointed out that China's wheat crop has declined by 8 percent since 1997, for example, and its rice crop by 4 percent—significant declines for the world's biggest producer of both grains.

Poor farming methods destroy soil structure and deplete currently used arable land of vital nutrients. Worse still, the world's virgin arable land is generally not as good as the land already under cultivation. In fact, NASA has found that the most productive land tends to sit below urban areas. People naturally picked the most fertile locations on which to build their cities, another testament to the historical connection between cities and agriculture.

Water scarcity poses an even bigger threat than land scarcity. Only about 3 percent of Earth's water is fresh, and most of that is in glaciers—at least for a few more years. We are consuming fresh water more quickly than it can be recharged by rain or snow, and about two-thirds of the fresh water we use goes to agriculture, mainly for irrigation.

Consider irrigation close to home. About 35 percent of irrigated cropland in the United States falls in two areas: the High Plains region and California's Central Valley. The High Plains region provides about 20 percent of US agricultural production (mostly grains) and is irrigated primarily with water from the High Plains aquifer, also the main source of local drinking water. The aquifer contains a rich supply of water trapped from the last Ice Age, so-called "fossil water." Fossil-water deposits are not rechargeable by rain, so once the aquifer is drained, it's drained for good. The High Plains aquifer is currently dropping about a yard per year. As farmers of old might have said, "We're eating our seed corn."

Similarly, the Central Valley of California—which reportedly produces about half the nation's fruits and vegetables in good times—relies heavily on irrigation; the area is nearly a desert without it. When in a severe drought, Central Valley farms face intense water competition that could result in an estimated 850,000 acres (343,983 hectares) going unplanted. This kind of situation could lead to a dust bowl. Just ask China, which is tapping fossil water and other aquifers at least as aggressively (sometimes lowering the water table below the reach of tree roots) and has already lost a California-size chunk of once-arable land to desert. India is expected to exhaust its water supplies by 2050. The same thing could happen to the High Plains aquifer by 2035 and to Lake Mead in the Southwest by 2021.

Bear in mind that China, India, and the United States are the three largest grain producers in the world. China and India may produce more by weight, but we produce considerably more per person. In fact, we have among the highest rates of productivity per farmer in the world, and that very likely goes a long way in explaining why urban agriculture has not blossomed in the United States: we simply have not needed it. Yet.

The *Other* Cuban Revolution

Urban agriculture flourishes in Cuba, where the vegetable and herb crop alone totals 4.2 million tons. Havana itself produces over half a million tons of food, all of it organic. Yet Cuba's evolution into an urban farming leader did not occur because of Havana locals' insistence on organic tomatoes wet with the morning's dew. It began ramping up in the early 1990s with the collapse of the Soviet Union, which had been providing two-thirds of Cuba's food, 98 percent of its oil, and most (if not all) of its synthetic fertilizers and pesticides. Cuba had to reinvent its entire food system, and it did so successfully within a decade.

Cuban agriculture in the 1980s had been very much like that of most industrialized nations: highly productive, highly mechanized, chemical-laden, and focused on a handful of crops, such as sugar and citrus. It could export these profitably to its main trading partner, the USSR, and import the majority of its food supply—until unforeseen political events turned off Cuba's oil tap and diminished its biggest market. Cuba suffered during the transition but now has a healthier, more resilient, and sustainable food supply involving thousands of organic urban farms, called *organipónicos*.

Our productivity rates also explain why many of the countries with well-developed urban agricultural programs are either big countries with large, historically hard-to-feed populations (such as China, Vietnam, and Brazil) or small countries with few natural resources (such as Singapore, Hong Kong, and the Netherlands). Necessity is the mother of invention.

The United States, like other countries, is not immune to the growing conflict between energy and water. The rising consumption of both leads to complicated, interdependent, and sometimes surprising results illustrated all over the world. In some Indian states, about half of the electricity goes to pump water up from ever-deeper tables underground, causing blackouts. Droughts and heat have caused the shutdown of Brazilian hydropower plants and European nuclear plants, which need vast quantities of water for cooling. The desalinization of seawater, often proposed to solve the water problem, requires a tremendous investment of energy, which helps explain its popularity in the water-poor, hydrocarbon-rich Middle East. The plug-in electric and biofuel-run cars that we hope will wean us off oil consume about ten to twenty times as much water per mile as gasoline (after all, you need water to grow the corn or sugarcane or other source of ethanol). According to Brown, "The grain required to fill a 25-gallon (95-liter) SUV tank with ethanol could feed one person for a year."

Joining food and water on the shrinking-planet front is their natural conclusion: waste. Lots of it. Population growth is a problem of inputs (shortages of food and water) and outputs (an abundance of waste). This ranges from gray water (e.g., dishwasher and washing machine effluent) and organic solid waste (e.g., food scraps and paper) to inorganic solid waste (metal, glass, and most plastics) and black water (you can guess). Americans produce about 4.5 pounds (2 kg) of solid waste per person per day, 70 percent of which is paper, food scraps, yard trimmings, and plastics. About a third of our solid wastes are somehow reused, by recycling or energy generation through incineration, but two-thirds just end up in landfills.

Landfills can pollute water through leaching and contribute to global warming through methane production and the fuel used in transport—not to mention that they just plain stink. Worse still, they're inefficient. Have you ever seen a movie in which a rich couple share a toast by a blazing fire and then throw their glasses to shatter in the hearth? Imagine that they do that two out of every three toasts. They

In addition to its other roles, the East Kolkata Wetlands provides important habitat for migratory birds and other animals.

would need to buy new glasses every other week or so—requiring mining of silica and huge energy expenditures for making glass—as well as find a place to dump their growing pile of shards. That's similar to what we do with everything. It doesn't make sense. (The actual US recycling rate for glass containers, incidentally, is about one in four.)

Wastewater management is no less important. In the developed world, wastewater is often treated at great cost and energy usage. In the developing world, wastewater is often untreated or treated in low-tech ways. Neither the typical developed-world model nor the typical developing-world model is ideal. The former throws away finite resources (water and macronutrients) that will have to be extracted from other sources—while at the same time polluting our biggest potential food source, the ocean—and the latter poses health risks. A middle ground is possible, however.

For example, Kolkata (formerly known as Calcutta) annually employs wetland-based ponds to treat much of its liquid and solid waste. The waste flows into pond-based wetlands, where it is cleaned by sunlight, bacterial action, and plants such as water hyacinth, which extracts toxic heavy metals. Every day, the East Kolkata Wetlands take in approximately 158 million gallons (more than 718 million liters) of wastewater and 2,500 tons of garbage, and they give back, on average, 30 tons of fish, 40 tons of rice, and 150 tons of fresh vegetables. Not a bad exchange. And farming, fishing, and other trades related to the wetlands provide about 30,000 full-time jobs. Many Americans would consider this reuse of water distasteful, but consider this: many older cities in the United States—such as New York, Philadelphia, and Milwaukee—have combined sewer systems. This means that sanitary sewage and stormwater runoff use the same pipe infrastructure. The virtue that inspired the system was constructing a single sewer system to handle two kinds of waste. Unfortunately, very high rainfall events (or sometimes even just average ones) can exceed system capacity, triggering a release of untreated sewage into local waterways, which closes beaches and can pollute local fisheries. Now *that's* disgusting.

In many parts of the world, food scraps, gray water, and other urbanite by-products are reused to create mulch, irrigation, and fertilizer. This is frequently the case in areas with the luxury of working sanitation systems or water-treatment plants but increasingly so in those areas where the recognition of waste's value has overcome an aversion to it. Israel and Jordan, respectively, reuse about 80 percent and 100 percent of treated wastewater for agriculture. Gray water is captured and reused in urban Jordan to grow olives, herbs, and other food—a practice made easy through simple plumbing modifications.

As population growth in general and poverty in particular are disproportionately the province of cities, it is becoming evident that cities should be the front line in addressing the triple challenge of food, water, and waste. "Urban farmers are not inherently more environmentally conscious than rural farmers," one FAO document notes. "They utilise urban waste because they farm the 2.5 percent of the earth where waste is most concentrated."

National Security

In 2009, the G8 nations convened a meeting of their agriculture ministers for the first time, acknowledging the critical role of agriculture. The ministers admitted that they were failing to reduce world hunger, a sentiment vindicated two months later when the FAO estimated the number of the world's chronically hungry to exceed 1 billion by the end of the year, an increase of 100 million in a year. "This is not just about food security," the *Financial Times* quoted USDA Secretary Tom Vilsack as saying at the G8 meeting. "This is about national security. It is about environmental security."

Urban agriculture should be a security priority for several reasons. The first and most obvious reason is to decentralize and strengthen our food supply, particularly in light of an increasingly global supply chain and centralized processing. In 2002, a prescient article by Peter Chalk in the *Rand Review*—a publication of the government-sponsored think tank—observed that "[m]ore by luck than design, the United States has not experienced a major agricultural or food-related disaster in recent memory. As a result, there is littler appreciation for either the threat or the potential consequences."

Just a year later, the first recorded instance of bovine spongiform encephalitis (mad cow disease) occurred in US cattle. No one was known to be infected, but it may be remembered as one of the first major incidents to pull the curtain away and reveal just what our food system had become. The incident revealed the difficulty of tracing back the cattle supply (to Canada, in this case) and exposed loose industry practices that allowed animal remains from livestock—reportedly including roadkill and even euthanized pets—to eventually wind up in the feed of some cattle, a herbivorous species.

In 2006, *E. coli* outbreaks resulting from lettuce and bagged spinach hit the nation, followed in 2007 by major outbreaks of salmonella (in pot pies, fresh spinach, and peanut butter), more *E. coli* (beef), and the first outbreak in decades of botulism in commercially canned products. Melamine-tainted products from China affected US pet food in 2007 and then, tragically, Chinese baby formula in 2008. Salsa ingredients, probably from both the United States and Mexico, were implicated in another salmonella outbreak in 2008, followed by a giant peanut-product-linked outbreak of salmonellosis in 2009.

Urban farming is not immune to such problems, but it could contain them locally and reduce the hazards posed by putting all of our agricultural eggs in just a few baskets. Consider spinach as the poster child against overconcentration of food production. Bagged spinach couldn't catch a break a few years ago, with *E. coli* contamination in 2006 and salmonella in 2007. Almost 75 percent of our spinach production occurs in California, meaning that a problem identified but not sourced in the California spinach supply could suddenly halt about three-quarters of the market in that vegetable. The outbreaks resulted in hardships on growers and their workers, a loss to Americans' diets, and bad news for Popeye.

Food shortages in the last few years have been even more worrisome. World commodity prices for core staples such as wheat, rice, and corn more than doubled from the end of 2006 to 2008, driven by a constellation of causes, including

droughts and storms, rising oil prices, biofuel production, agricultural subsidies, and a growing demand for meat. Higher food prices and the recession raised the number of Americans receiving Supplemental Nutrition Assistance Program (food stamp) funds by the end of 2008 to more than one in ten, while two hundred food banks served by Feeding America (formerly America's Second Harvest) registered an average 30 percent increase in demand. Riots concerning food prices erupted in roughly thirty countries.

More troubling than the sufficiency and integrity of our domestic food supply—which are rather troubling—are the destabilizing effects of shortages of food, water, and energy in the rest of the world. Secretary Vilsack bluntly described the possible scenarios posed by these hungry masses: "People could riot—that they have done; people migrate to places where there is food, which creates additional challenges; or people die." This has not been completely lost on the national security community, though the issue is mostly couched in terms of "climate change."

Three major reports on climate change and national security were released in 2007, commissioned by the government or produced by government-supported think tanks.

One report prepared for the military by the CNA Corporation aptly describes climate change as a "threat multiplier." Yes, it will worsen things, but the fundamental threats are already in play: increasing populations, decreasing fresh water and oil resources, degrading ecosystems due to overuse and pollution, and stagnating agricultural productivity. "Five billion people are expected to live in water-stressed countries by 2025," the report notes, "even without factoring in climate change."

The fundamental premise is that, already in many places in the world, water resources are stretched to the limit, people are flooding into cities without the service infrastructure to support them, and economies are precariously subject to the vagaries of a global market. These nations are already near "carrying capacity," a euphemistic term for how many people can be supported sustainably. "In such places," notes a report by the Global Business Network (GBN), "with such multiply stressed systems, the freak storm or prolonged drought could very well launch a profound, cascading crisis." It can end in a "failed state." Often, environmentalist Lester R. Brown points out:

> **It is not the concentration of power but its absence that puts us at risk.**
>
> **States fail when national governments can no longer provide personal security, food security, and basic social services such as education and health care. They often lose control of part or all of their territory. When governments lose their monopoly on power, law and order begin to disintegrate.**
>
> **Failing states are of international concern because they are a source of terrorists, drugs, weapons, and refugees, threatening political stability everywhere.**

What does this have to do with urban farming? A lot. Cities' dependence on imported food makes them extraordinarily fragile. They are subject to the whims of drought and flood in food-producing rural areas, to unexpected storms and political blockades, and to the price of the oil needed to produce and ship food to them. When governments fail to mitigate crises, notes the GBN report, it "reduces political legitimacy and halts economic activity, thus driving local populations to rely upon primary loyalties (families, neighborhoods, religious organizations, gangs, and so on) for daily survival. This dynamic in the political system is often (and will increasingly be) played out in urban settings...." The GBN cites Hurricane Katrina as an example of a government's failure to deliver—if that's how the most powerful nation responds to crisis, what can we expect of the weakest fifty or one hundred?

The people who thrive in failed states are the black marketers, pirates, drug smugglers, terrorists, and others who benefit from chaos. That's a huge threat. Equally dangerous, however, is the honest majority of people who just want to get out. A conservative estimate is that there are 20 million environmental refugees right now—all of whom were directly forced from their homes by the environment or through its geopolitical effects—and there will be at least 50 million by 2020; many estimates are significantly higher. Hypothetically speaking, if the United States took in a share of those 30 million additional refugees proportional to its 4.5 percent share of the world population, that would mean an extra 1.3 million residents.

Just think of it in terms of individuals instead of nations. If yours is the only thriving farm in a valley stricken by drought, to whose farm will all the peaceable starving people come begging for food? And how about the angry starving people with pitchforks and torches? While market-based considerations make it desirable for

your farm to do better than other farms, you have an even greater interest in the other farms doing well enough that they have resources to trade and don't turn into crowds of refugees or criminals at your doorstep.

An organipónico in the same downtown neighborhood as the United States Interests Section.

At the metropolitan or national scale, urban farming provides some insurance against situations getting that desperate. Urban farming can also produce food with a fraction of the fossil fuels that go into traditional agriculture. Perhaps even more importantly, it can do so with less fresh water, particularly if it's integrated with a city's wastewater plan.

The fact is, agriculture (mainly through field irrigation) sucks up close to 70 percent of the freshwater used by the world every year, increasing the risk of water wars that many experts fear. Many of the largest lakes and rivers in the world border more than one country. (The United States alone contains over sixty river borders.) If one country restricts, diverts, or pollutes the water, it could have dire effects on—and prompt desperate measures by—its neighbor(s). Some of the places where the risks are greatest are already among the most volatile regions of the world, such as the Middle East and Central Asia.

But, once again, we need look no further than home. Thanks to a drought in the Southeast in 2007, Georgia politicians eyed moving their state's border to access the Tennessee River—ostensibly to correct an old surveyor's error—but have since backed down...for now. Georgia was, in turn, sued by Alabama and Tennessee to halt its plans to access the water of Lake Lanier, which sits in Georgia but feeds the Chattahoochee River vital to the other two states. South Carolina and North Carolina wrangled over access to the Catawba River's water. And these are all relatively water-rich Eastern states; freshwater is much scarcer in the West, where it has long fueled conflicts.

If these issues can bring conflict—if not violent, at least litigious—to states within one of the most stable, water-rich nations, just imagine how they will affect nations that are not so friendly with each other. China controls the Tibetan Plateau, for example, which houses not only the headwaters of its own main rivers—the Yellow and the Yangtze—but also those vital to its southern neighbors, including the Mekong (Burma, Cambodia, Laos, Thailand, and Vietnam), the Brahmaputra (Bangladesh and India), and the Salween (Burma and Thailand). Diminishing glaciers, snowpack, and permafrost threaten the flows of all the rivers, and pollution increasingly taints the water that is there. These rivers provide not only a means of transport but also drinking water, irrigation, food, and a main source of electricity. The same can be said of rivers and other transboundary waters worldwide.

Urban farming is an obvious choice to mitigate both the threat of food-supply chains whose ultimate sources we cannot identify and the possibility of the disruption of imported food supplies. It is no accident that urban farming has cropped up in places beset by war, disaster, poverty, or instability: Sarajevo under siege, post-Communist Russia, Rwanda after civil war, Honduras after Hurricane Mitch, Kirkuk after the toppling of Saddam Hussein, and so on. Urban agriculture is an effective (if not comprehensive) response to such crises but is even better as preparation for them.

China started an urban agriculture strategy in the 1960s to prevent once-common urban famines. Almost twenty major Chinese cities are now self-reliant in foods other than grain. No urban famine has struck China since World War II, and no famine of any kind has occurred since the 1960s. The urban farms Cuba built due to the political and economic crisis of the Soviet Union's fall served it well when three hurricanes raked the country in 2008. Fast-maturing crops grown in urban areas helped quickly get food to the populace. These are not perfect systems, but they are functioning and largely effective, developed by countries addressing their vulnerabilities programmatically at the national level, head on. We would do well to do so ourselves, and—in an increasingly interdependent world—to help other nations do so, too.

Most of the organizations currently promoting urban agriculture are based outside the United States and focus their work there, but there is a groundswell of people poised to gather the political will to accelerate this work within our own country.

The Future: Fungible Goods...and Bads

In 1987—before the USSR collapsed, before the creation of the World Trade Organization, even before the World Wide Web—the United Nations' World Commission on Environment and Development released a report titled "Our Common

Future" that starkly quantified the population challenges facing our world: "Our human world of 5 billion must make room in a finite environment for another human world.... More than 90 percent of the increase will occur in the poorest countries, and 90 percent of that growth in already bursting cities." Even more prophetic, perhaps, was the text immediately preceding it:

> **Until recently, the planet was a large world in which human activities and their effects were neatly compartmentalized within nations, within sectors (energy, agriculture, trade), and within broad areas of concern (environment, economics, social). These compartments have begun to dissolve. This applies in particular to the various global "crises" that have seized public concern, particularly over the past decade. These are not separate crises: an environmental crisis, a development crisis, an energy crisis. They are all one. The planet is passing through a period of dramatic growth and fundamental change.**

The planet has changed, and many feel not for the better.

A common objection to our economic machine from many quarters is the indignity—even absurdity—of a market that assigns value to things believed to be rightly immeasurable and thus outside the market. Author, poet, and farmer Wendell Berry writes,

> **The way of industrialism is the way of the machine. To the industrial mind, a machine is not merely an instrument for doing work or amusing ourselves or making war; it is an explanation of the world and of life. Because industrialism cannot understand living things except as machines, and can grant them no value that is not utilitarian, it conceives of farming and forestry as forms of mining; it cannot use the land without abusing it.**

In his allusion to mining, Berry echoes many proponents of sustainability (not just in agriculture) who think of the current state of affairs as being extractive—of natural resources, of people, of heritages being lost from our past, and of gifts being robbed from the future. The extraction of natural resources is obvious, from oil to ebony to lobsters. The other extractions—such as self-reliance and a connection

with nature—are harder to quantify, precisely because they are nonmarket values. And the course of action is so hard to plot—I think, in part, because the circumstances we live in are so alien to most of human experience.

Never before have average people wielded so much power with so little skill. If you want a shed, with the click of a mouse you can direct one to be built from materials you've never seen, by people you don't know, and in a country you've never visited and, perhaps, cannot even pronounce. It might then be shipped to another country for preassembly before being delivered to your doorstep. You don't need to know how to hammer a nail, plumb a line, or cajole a neighbor. Unlike our ancestors, who acquired and employed an array of skills to survive, our abilities tend to be either so general as to make us interchangeable or so specific as to make us helpless outside our professional bailiwicks.

In fact, one skill we're losing, according to thinkers such as environmentalist and author of *The End of Nature* (2006) Bill McKibben, is that of being a neighbor. "A meteorite could fall on your cul-de-sac tomorrow, disappearing your neighbors," he writes, "and the routines of your daily life wouldn't change." We no longer need each other in the kind of direct way our ancestors did. Oh, sure, we need someone somewhere for something, but the who, where, and what don't matter so much anymore. It's become cliché to observe that even as our global neighbors become closer, our local ones seem ever more distant.

We face debilitating diet-related morbidities not out of scarcity but out of abundance. The amount Americans overeat could supply the entire caloric needs of Bangladesh, of France and Germany combined, or of the total populations of North Korea, Yemen, Taiwan, Ghana, Malaysia, and Venuzuela plus 10 million of their closest friends. The problem isn't just how much we eat, but what we eat: highly processed foods from a homogenized selection of choices—the very kinds of foods made possible, if not inevitable, by industrial agriculture.

It should come as no surprise that urban farming addresses all of these issues. One of the most common reasons for starting farms is not primarily to grow food but to learn skills: to teach entrepreneurship to women in Africa, for example, or to develop leadership among youth in the United States. Community-building is another key function of urban agriculture; in fact, it's maybe the most common one for the 20,000-plus community gardens in the United States and Canada.

There are many motivations for urban farming. Besides those just described, there is the enduring allure of combining city life with nature's beauty—the best of both worlds. And what is more basic than food? As H.C. Flores writes in *Food Not Lawns*, "[G]rowing food is one of the most radical things you can do: Those who control our food control our lives, and when we take that control back into our own hands, we empower ourselves toward autonomy, self-reliance, and true freedom." Since a majority of people now live in urban areas, the only avenue for growing food for most of us will be some form of urban farming.

Beneath it all is a desire to be grounded. And we will indeed need our wits about us if we are to attempt large-scale urban agriculture, whose role in the sustainable-farming movement is described in chapter 2. (Of course, if you just want to roll up your sleeves and get started, you can skip ahead to Part II: Your Own Backyard.)

CHAPTER 2

Marching to Sustainability on Our Stomachs

Urban farming as a concept strives for sustainability. It's clear what sustainability is not: the status quo. What sustainability is...well, that's more complicated. Rooted in the recognition that our current course of action cannot continue (and often in the belief that it should not continue), sustainability offers no simple, universally accepted alternative. So it is the broader sustainable agriculture movement in which urban farming fits most comfortably. As noted by Mary V. Gold and Jane Potter Gates, who have traced the evolution of organic and sustainable agriculture, "'Doing' sustainable agriculture is a less complicated endeavor than defining it." Let's take a look.

Sustainable Agriculture

The National Sustainable Agriculture Information Service (ATTRA) defines sustainable agriculture as "agriculture that follows the

principles of nature to develop systems for raising crops and livestock that are, like nature, self-sustaining. [It] is also the agriculture of social values, one whose success is indistinguishable from vibrant rural communities, rich lives for families on the farms, and wholesome food for everyone."

If that's a little too touchy-feely for you, fear not! Congress has taken it upon itself to define sustainable agriculture. According to the 1990 Farm Bill:

> **The term "sustainable agriculture" means an integrated system of plant and animal production practices having a site-specific application that will, over the long-term:**
>
> **(A) satisfy human food and fiber needs;**
>
> **(B) enhance environmental quality and the natural resource base upon which the agriculture economy depends;**
>
> **(C) make the most efficient use of nonrenewable resources and on-farm resources and integrate, where appropriate, natural biological cycles and controls;**
>
> **(D) sustain the economic viability of farm operations; and**
>
> **(E) enhance the quality of life for farmers and society as a whole.**

According to ATTRA, sustainable agriculture's "midwives [are] not government policy makers but small farmers, environmentalists, and a persistent cadre of agricultural scientists. These people saw the devastation that late twentieth-century farming was causing to the very means of agricultural production—the water and soil—and so began a search for better ways to farm, an exploration that continues to this day." It is these people who have given rise to sustainable methods of production and consumption.

Externalities

Critics often refer to the "externalized costs" of industrial agriculture. These are negative effects caused by an economic decision (spending or saving), but borne primarily by outsiders to the transaction. A hypothetical scenario makes the problem clear. Let's say, for example, that a town assesses an annual fee of $200 for trash-collection services. It's technically voluntary, but who wouldn't want his trash hauled away? Let's say a guy named Tom.

Pennypincher that Tom is, he refuses the service and starts simply throwing garbage out the kitchen window. When the heap gets really high, he uses the window of the room directly above the kitchen.

The dump attracts all manner of vermin. Before you know it, rats' nests in chimneys pose fire hazards for a ten-block radius, downwind property values have plummeted, and little Sally across the street has typhoid fever. Yet Tom, the economically rational actor, has saved $200.

The true costs of Tom's economic action have been foisted upon (externalized to) his neighbors and would almost certainly exceed $200 if they could be valued accurately. Yet causation is hard to prove, redress is expensive to obtain, and the actual costs are difficult to tally. Sure, the decline in property values and treatment of Typhoid Sally might be quantifiable. But what is the cost, for example, of neighbors fearing that they smell like trash heaps when they go to work?

It's clearly some cost. If you genuinely fear that your clothes have absorbed the smell of your crazy neighbor's mini-landfill (you can't tell for sure, because the smell pervades your house), what would you be willing to pay to make sure that you're odor-free? The cost of an extra dry cleaning?

According to critics, it is externalized costs and generous subsidies that keep industrial agriculture going. Food is cheap—far cheaper as a percentage of one's income than in the past—but this cheapness comes at great cost. Annual subsidies range from about $10 to 25 billion, depending upon the year, and economists Erin M. Tegtmeier and Michael D. Duffy estimate the externalized costs at $5.7 to 6.9 billion annually. They also believe that to be a conservative estimate, having calculated only the fairly tangible costs, such as treatment of common foodborne illnesses and pesticide poisonings in humans, treatment of polluted waters, and putting regulations in place.

Sustainable Production

If, as many people believe, full-scale industrial agriculture constitutes an unsustainable way of producing food, then the first challenge of sustainability concerns finding adjuncts, if not alternatives. There are many claimants to the throne of sustainability.

Biodynamics and Organic Farming

Biodynamics has its roots in the early 1920s in Germany, when, even then, farmers noticed the decline in their soil quality. Some entreated Rudolf Steiner, a philosopher and founder of anthroposophy (a spiritual philosophy and practice), to help them address the degradation of the land. It may seem strange to modern Americans weaned on technology that the farmers would first approach someone affiliated with esoterica for a solution, but he was a prominent, well-educated man familiar with traditional peasant farmers displaced by the rise of early industrial agriculture. A subsequent series of lectures by Steiner produced, according to the Biodynamic Farming and Gardening Association, the "fundamental principles of biodynamic farming and gardening, a unified approach to agriculture that relates the ecology of the earth-organism to that of the entire cosmos."

Steiner viewed the farm itself as a kind of organism, and the then-developing class of chemical inputs—synthetic fertilizer, herbicides, and pesticides—as detrimental to the soil and the farm. Central to the biodynamic approach are the concepts of working with nature's rhythms (including astrological considerations), crop rotation, emphasis on manures and compost, a handful of special numbered "preparations" to facilitate beneficial actions such as compost formation or root growth, and cultivation of one's inner self along with the land.

Demeter International has certified biodynamic farms since the late 1920s, and biodynamic certification requires full organic certification plus meeting additional standards; in fact, the entire farm, not just a particular crop or field, must be certified. As agribusiness increases its share of the organic food market and some sustainable consumers look for something "beyond organic," biodynamics is a frequently mentioned candidate.

What we now call organic farming originated, to a large extent, in biodynamics. In fact, Lord Northbourne, a Steiner devotee who embraced the biodynamics idea

Inputs and Outputs

Modern agriculture is often discussed in terms of inputs and outputs. Inputs—such as water, light, seeds, fertilizer, herbicides, or pesticides—provide the means of realizing the desired outputs, such as food, fiber, or lumber. Some outputs limit input options. A certified organic apple (output), for example, would exclude genetically modified seeds, restrict allowable fertilizers, and permit only certain kinds of herbicides and pesticides or methods of achieving the same end (for example, integrated pest management).

In more conceptual terms, inputs of agriculture include labor, capital, and land. Land is typically the top limiting factor for urban farmers, followed by capital to buy seeds, fertilizers, and other material inputs. There are also inputs and outputs not often considered. For example, oil and natural gas is often an implicit agricultural input—to power equipment, operate refrigerators, transport goods, and synthesize agrochemicals—while pollution caused by runoff of agrochemicals or their nonsynthetic equivalents is often an output. In fact, the ostensible threat of pollution is one of the leading objections that municipalities make to urban agriculture.

Pollution falls short as an objection from both input and output perspectives, however. Yes, cities can be polluted, but urban farmers often use raised beds or containers with clean, uncontaminated soil and make use of composts that mitigate the effect of many pollutants. In addition, rural growing areas are often as polluted as urban ones, if not more so. So the pollution-as-an-input argument lacks bite. The pollution-as-an-output argument fails to impress, as well, since urban farmers usually tend crops closely by hand and grow organically or nearly organically. You're unlikely to find them deploying crop dusters with pesticides, lugging half-ton fertilizer spreaders behind a tractor, or emptying thousands of gallons of pig waste into manure lagoons.

of the farm as a complete living entity—an organism—coined the term in 1940. In the United States, Jerome Irving Rodale, who was a dedicated farmer and successful publisher, became the face of organic farming in this country. His 1948 classic *The Organic Front* combined the research of early pioneers such as Lord Northbourne with his own experimental findings.

The International Federation of Organic Agriculture Movements (IFOAM), established in 1972, defines organic agriculture as "a production system that sustains the health of soils, ecosystems, and people. It relies on ecological processes, biodiversity, and cycles adapted to local conditions rather than the use of inputs with adverse effects. Organic agriculture combines tradition, innovation, and science to benefit the shared environment and promote fair relationships and a good quality of life for all involved."

From a consumer standpoint, and from the standpoint of a producer looking for the all-important "certified organic" label, it is, in effect, a nebulous philosophical movement that has been reduced to a precise set of codes and practices. "It is a process claim," notes the FAO, "rather than a product claim." In other words, it's a *how* rather than a *what*. It says nothing about an organic apple being healthier, fresher, tastier, or otherwise better than a conventionally grown one, for example. It says only that the certified apple was grown a certain way. The consumer can

Oh, Sludge!

While the White House Kitchen Garden quickly became a sensation, one thing it could not become was certified organic. As Michelle Obama learned, USDA organic standards forbid the use of sludge, which had been applied to the grounds during a previous administration. Sludge is defined as "[a] solid, semisolid, or liquid residue generated during the treatment of domestic sewage in a treatment works." This prohibition might seem to contradict the whole organic ethic and its devotion to rebuilding soil quality through nonsynthetic inputs and practices; in fact, sludge might well have been used at the White House as an environmentally friendly alternative to synthetic fertilizers. And, after all, organic farmers use animal manure, so why not, you know, human? The reason is that much of our sewage combines residential and industrial waste, resulting in contamination by substances such as mercury, lead, and dioxins. That's the crap they're worried about.

Two of these peaches are organic, and one is conventionally grown. Can you tell the difference?

draw his or her own conclusions about the other qualities.

So what are the production restrictions of certified organic food? The specifics—and be sure, there are a lot of them—vary by jurisdiction but generally involve three principles: use of natural inputs (such as fertilizers or pesticides) rather than synthetic ones, no use of genetically modified organisms, and practices resulting in soil enhancement rather than depletion. There are exceptions, though. An FAO report notes, for example, that arsenic would be prohibited as an insecticide even though it is natural (and it is banned in the feed for USDA-certified organic chickens), and synthetically made insect pheromones might be acceptable for trapping harmful insects organically.

An organic peach may look the same, taste the same, and be nutritionally identical to a conventional peach. But all of these qualities are subject to vigorous debate, and that, in fact, is why organic certification is necessary. The consumer cannot confidently distinguish organic from conventional (other than by the 20 percent markup in price). Again, the organic label speaks to production methods, not food qualities. It does not even guarantee the absence of synthetic pesticides (since these pervade our environment), although studies repeatedly show much lower levels of these chemicals in organic food. Just as organic produce purposely grown without any synthetic pesticides might contain them because of environmental contamination, some "genetic pollution" from genetically engineered crops may also be acceptable. The European Union, for example, allows organic food to contain up to 0.9 percent genetically modified content, citing "adventitious or technically unavoidable" contamination.

It's worth noting that biodynamics and the early organic farming movement began at the tail ends of World Wars I and II, respectively, in the industrialized nations involved in those conflicts. This happened in large part, no doubt, because of the rapid appearance—and then disappearance—of fertilizers. Germany, the United Kingdom, and the United States all took early to industrial agriculture, which is highly dependent on nitrogen fertilizers. The same sources of fertilizer were also

Frankenstein for Fruit

What does "genetically modified" (or "genetically engineered" or "GMO") mean? It means that something that has been directly manipulated at the genetic level to possess (or lack) certain characteristics. This contrasts with traditional plant and animal improvement, which relies on selectively breeding organisms with the features one wants. Let's say two tomato plants out of hundreds survive an outbreak of blight, so you plant the seeds of those two tomatoes. Blight hits again, and ten plants survive. Now you plant the seeds of those ten tomatoes. After a few generations of repeating this process, you will (ideally) end up with a strain of tomato uniformly resistant to blight.

Genetic engineering, however, takes a more direct approach. It might involve snipping out some genes that make the average tomato susceptible to blight, or splicing in some genes from a related plant resistant to blight. Or it might be even more exotic. A majority of the corn grown in the United States has genes from the common soil bacterium *Bacillus thuringiensis* (*Bt*) spliced into it. This bacterium has the useful quality of killing all kinds of insect pests (but not humans) that eat it; for that reason, it has long been sprayed on crops. Then someone thought: instead of taking all the trouble to culture *Bt* in the lab and then spray it on the crops—where it might miss some spots or get washed off in the rain— how about we just take the genes from *Bt* that kill pests and insert them into corn, so the corn would manufacture its own pesticide? And that's exactly what has happened. Most of our cotton also includes *Bt* genes, while the majority of the soybean and sugar beet crops (and some corn) contain another modification that makes them resistant to the herbicide glyphosate so that entire fields can be sprayed with glyphosate from above, letting the genes sort out weed from crop.

needed to make explosives—even today, we worry about fertilizer bombs such as the one used in Oklahoma City—so wartime shortages of fertilizer revealed just how dependent upon chemical inputs agricultural land had become and how impoverished it was without them. The result was renewed interest in older forms of agriculture and in the exploration of new ones.

Certified biodynamic and organic farming both require detailed record keeping and are particularly challenging for urban farmers because of land-tenure issues. Many urban farmers do not own the land they farm, and the areas farmed may shift every few years as the farmers move, vacant lots currently used for urban farming get developed, new sites become available, and so on. Biodynamic and organic certification both require several years without forbidden inputs, which may be hard for urban farmers to achieve or establish.

Know Your Numbers

All of us have made that most grievous of supermarket sins: bringing unlabeled produce to the checkout line. Those price lookup (PLU) codes identify the kind of product, letting the cashier distinguish between, say, cilantro (4889) and Italian parsley (4901), which look quite similar. Organic produce has a five-digit code that starts with a 9; organic cilantro, for example, would be 94889. Genetically modified produce (which, by definition, cannot be organic) has a five-digit code that starts with an 8, such as 84901 for genetically modified Italian parsley. So, to recap: four digits mean conventional produce, five digits starting with a 9 signify organic produce, and five digits starting with an 8 identify genetically modified produce.

Other Farming Methods

Intensive farming is not sustainable or unsustainable, per se; it all depends upon how the intensification is achieved. Unsustainable methods might include repeated plantings of the same crop in the same space, heavy application of synthetic fertilizers and pesticides to make up for the soil's declining health, or densely populated aquaculture without any means of recycling and cleaning the water. Intensive farming often refers to sustainable intensification methods, such as intercropping or succession planting. The most familiar model of sustainable intensive farming is the "biointensive" method associated with John Jeavons and Ecology Action, for whom an early influence had been Alan Chadwick, a student of biodynamic farming and an inspirational figure in the organic movement.

Ecology Action's "GROW BIOINTENSIVE" method involves eight main principles:

1. using raised beds that are double-dug, providing a full 24 inches of good tilth;
2. composting;
3. "carbon farming" with crops such as corn or oats, which provide both calorie-dense food and carbonaceous ingredients for composting;
4. intensive planting (to maximize yield and shade out weeds);
5. companion planting;
6. calorie farming (growing crops that provide a lot of food bang for the space they use);
7. using open-pollinated seeds, which preserve diversity; and
8. employing a "whole-system farming method," which integrates all of these methods to ensure that the soil is replenished at a rate consistent with its high production.

SPIN-Farming® (SPIN stands for "small plot intensive") likewise encourages certain high-productivity techniques but is distinguished from other farming methods mainly by its savvy and unapologetic focus on profitability for small farmers. Focusing on farmsteads of less than an acre, SPIN-Farming presents itself as the closest thing to a franchise you can find in small-scale agriculture. The system can integrate just about any farming technique or philosophy you might want.

At the opposite end of the effort spectrum is natural farming—sometimes called "do nothing" farming—a fundamentally organic approach pioneered by Japanese scientist and farming visionary Masanobu Fukuoka. In fact, Fukuoka is often cited as a major influence in the organic movement. Natural farming encourages such practices as breaking up compacted soil by using deep-rooted vegetables rather than by tilling, casting seeds on the ground rather than planting them, and employing soil-improving plants rather than chemical fertilizers, favoring integrated pest management over artificial pesticides, and being generally unconcerned with weeds.

No-till farming avoids tilling the soil by using special tools to plant seeds in the vegetative stubble of the most recent harvest. This method saves time, usually

Architect Vincent Callebaut's vision for a dragonfly-inspired "metabolic farm" on New York City's Governors Island, viewed from the Empire State Building.

improves drainage, and dramatically reduces soil erosion and evaporative water loss. It also helps sequester atmospheric carbon and saves the 3.5-plus gallons (13.25 liters) of fuel that would normally be used to till an acre. *No-dig gardening* minimizes excavation of the soil by building upward with mulches.

Realizing that three dimensions offer a much greater growing area than just two, farmers have been farming vertically for centuries, from simple beanpoles to the Hanging Gardens of Babylon, and from ancient terraced fields in the Philippines and Peru to the trellises and trees used by the ancient Romans to support growing grapevines. As the term is most commonly used now, however, *vertical farming* refers to skyscraper farms built for the purpose.

Imagine a thirty-story transparent building—electrically self-sufficient through renewable sources—that grows enough food for 50,000 people. The leading proponent of this idea is Columbia University professor and author of *The Vertical Farm*, Dickson Despommier, whose ideas have ignited the imaginations of architects all across the world (see chapter 9). Many of these architects' designs seem whimsical, but they are no more fanciful than the prospect of continuing with the status quo.

Hobby farming describes the practice of "traditional" smallholder farming and related activities (handicrafts and so on) in a rural setting for recreational purposes or as a sideline business. It is the counterpart to recreational farming in urban agriculture. *Homesteading* is almost the complete opposite: a life in which farming and self-sufficiency are vocation rather than avocation.

Specialized Farming of Trees, Shrubs, and Other Perennials

Most of us are familiar with relatively unmanaged woods, ornamental groves, and orchards, though forests have much broader uses. *Agroforestry* is an agricultural approach to forestry, usually conceived as including uses beyond timber and greater diversity than a single species of tree or even class of trees. Planting 300 acres (121.5 hectares) of white oak for lumber would not quite cut it as agroforestry. Mixing in sugar maples; native beeches, pecans, and hickories; understory serviceberries and native persimmons; and shiitake mushrooms probably would. All of the overstory trees and the persimmons would provide excellent hardwoods. Some fresh oak, pecan, and beech logs could be kept in place to cultivate shiitake mushrooms. The maples could provide syrup. The pecans could provide nuts, and the persimmons and serviceberries could provide fruit. Pigs could forage on acorns, beechnuts, hickory nuts, and stray pecans. This agroforestry approach could be equally profitable while also being more sustainable.

Forests can be used as living fences in the human-defined landscape, including as windbreaks and riparian buffer strips, which help prevent erosion and protect water resources from airborne pollution. Livestock, such as cattle, can be combined profitably with forest in the practice of *silviopasture*, where strips of open land are maintained for the animals in between stretches of trees. The cattle can keep grass and weeds in check while providing high-quality, low-cost fertilizer. Sometimes rows of vegetable crops (or even edible bush crops) are alternated with rows of trees in a method called *alley cropping*. Growing alleys of wheat between nut trees would be an example akin to cultivating radishes or other "catch crops" (rapidly growing crops making use of temporarily available space and resources) amid tomatoes or corn. The function is similar, too: smaller crops making the most of unused resources (sunlight, water, nutrients) while the larger main crops are yet to mature.

Forests provide the natural habitat for many nuts and berries as well as for other culinary or medicinal plants and fungi. In fact, two of the world's most expensive gourmet items, truffles and saffron, originate in forests: temperate hardwood and arid shrubland, respectively. Bees can produce honey from tree blossoms, which include culinary favorites such as tupelo, eucalyptus, fir, sourwood, citrus, and basswood.

A good example of urban agroforestry is "Mi Programa Verde," an initiative in Havana to plant 17 million trees, all of them with secondary uses. Ranging from woods-level plantings to patio trees, the program aims to provide the city with a permanent supply of tree fruit, wood fuel, and timber. This is dwarfed by the ambitious but problematic "Green Wall of China," that country's initiative to plant a 2,800-mile (4,506-km) wooded strip to stop further loss of arable land by keeping the Gobi Desert at bay.

Another approach suited to larger urban areas would be edible forest gardens. This concept is closely associated with David Jacke and Eric Toensmeier, who wrote a book of the same name. Jacke and Toensmeier encourage an approach deeply informed by ecology. It is "farming like the forest," as they put it, rather than merely "farming in the forest," as might be done in silviopasture or other simple agroforestry projects.

Trees and shrubs are also dear to the heart of *permaculture*, a word embracing both "permanent agriculture" and "permanent culture" that was first detailed in the book *Permaculture One: A Perennial Agricultural System for Human Settlements* (1978) by Bill Mollison (who coined the term) and David Holmgren. In fact, a few early books had hit upon the same goal, if by different means, including *Soil Fertility and Permanent Agriculture* by Cyril George Hopkins (1910); *Farmers of Forty Centuries or Permanent Agriculture in China, Korea, and Japan* by Franklin Hiram King (1911); and *Tree Crops: A Permanent Agriculture* by Joseph Russell Smith (1929).

Like natural farming and edible forest gardening, permaculture strongly emphasizes working with nature rather than against it. "The aim of permaculture," according to Toby Hemenway, author of *Gaia's Garden* (2001), "is to design ecologically sound, economically prosperous human communities. It is guided by a set of ethics: caring for Earth, caring for people, and reinvesting the surplus that this care will create. From these ethics stem a set of design guidelines or principles, described in many places and in slightly varying form."

The design principles are too detailed to discuss here, but the resulting designs have several distinguishing features, including diverse polyculture, a partiality toward plants with many uses (e.g., bamboo, hemp, coconut, or amaranth), multiple layers of vegetation (tree, vine, shrub, herb, and so on), nonlinear borders, and an emphasis on perennial food plants (such as apples, blueberries, or asparagus) over the annual plants (such as corn, tomatoes, or broccoli) at the center stage of most agriculture.

Edible landscaping is exactly what it sounds like: landscaping with food plants. Although edible forest gardens or permacultural gardens might technically fit the category, the term is usually used to connote landscaping that is ostensibly traditional in form and function—hedges and flower beds, for example—but that also provides food. Elderberry or blueberry shrubs might fill in for a privet hedge, for example, and an actinidia (kiwi) vine might substitute for Virginia creeper. Gardening beds might contain edible flowers, such as nasturtium and borage, among annuals, and camassia, day lilies, and alliums among perennials. Part of the appeal is the quest to find different edible plants to replace inedible ones with similar cultural requirements.

Edible schoolyards are gardens (often at or near schools) that combine hands-on work

Many species of the genus *Mahonia* make excellent subjects for edible landscaping.

A 20,000-square-foot green roof above Chicago's City Hall built to test the benefits of the concept. Chicago is a leader in green roofs, with nearly 6 million square feet (557,400 sq m) of roof space devoted to them.

by students with curricular requirements. The most famous one—and probably the first of its kind—is the Edible Schoolyard in Berkeley, California, that is associated with chef, writer, and food advocate Alice Waters. Begun in 1995, this project and its affiliates also integrate the schoolyard closely with the food program of Martin Luther King, Jr. Middle School.

Green Roofs

Starting in Germany in the 1960s, green roofs have taken off in the developed world as a means to provide an array of benefits in the built environment, aesthetics being only one. Water management is probably the biggest selling point in cities. Green roofs can absorb up to 90 percent of rainwater, releasing it eventually through evaporation or a gradual flow into the sewer system. This is no small feat: a midsize city building with a flat 100-by-150-foot (30.5-by-48-m) roof will shed more than

Sedum

18,000 gallons (68,137 liters) of water during a rainfall of just 2 inches (5 cm)—and there may be ten or more buildings of that size on a single block.

Green roofs also help to moderate the urban heat-island effect, which is the tendency of some cities to stay several degrees warmer than surrounding undeveloped areas, particularly at night. They provide insulation for the buildings hosting them, reducing both heating and cooling costs. They also help dampen noise, filter air, and provide habitat for wildlife, such as birds.

Intensive green roofs are those with a significant soil depth that can accommodate larger plants and that often require more hands-on maintenance. The classic rooftop garden with trees, bushes, and maybe some flowers would be intensive. *Extensive* green roofs, in contrast, generally have a very shallow planting depth—maybe just a few inches—and are designed to be very low maintenance. They are more popular for that reason and often are populated with grasses and low-growing native plants.

By far, the most popular plants for urban green roofs have traditionally been various species of sedum (pictured). Hardy succulents, sedums thrive on neglect, withstand full sun and drought, and often stay low enough to be relatively unaffected by wind. Yet sedum plantings are also very much like the shag carpets of the plant world: high-piled, gaudy, and so very 1970s.

Most sedum is inedible, some varieties are toxic, and wildlife makes little use of it. A newer trend has been to design green roofs to include food crops and other functional plants.

Community and Allotment Gardens

Community gardens are units of land shared by local residents, who often have their own small plots within them. They are similar to *allotment gardens* in Europe, which began during the Industrial Revolution to help city workers feed themselves. The American Community Gardening Association (ACGA) estimates that there are about 18,000 community gardens throughout the United States and Canada, many of which are listed in ACGA's online database (https://communitygarden.org/find-a-garden).

Community gardens are a major means of urban farming in the United States, and the benefits that ACGA attributes to these gardens echo those of urban farming in general: "improv[ing] people's quality of life by providing a catalyst for neighborhood and community development, stimulating social interaction, encouraging self-reliance, beautifying neighborhoods, producing nutritious food, reducing family food budgets, conserving resources, and creating opportunities for recreation, exercise, therapy, and education."

Growing Techniques

Container gardening is just that: cultivating plants above ground in something that holds a reasonable volume of soil—from an empty tin can to a large planter to a column of old tires. *Windowsill gardening* is simply container gardening that uses window boxes or other receptacles by a window.

Raised-bed gardening employs an aboveground, level volume of soil for cultivation, usually within some kind of frame. It is a popular method of small-scale agriculture the world over for its many advantages. Raised-bed gardening allows for workable, productive soil on top of severely compacted ground, or sometimes even asphalt. Where native soil is contaminated, it provides a space to grow food in clean soil with minimal uptake of contaminants. In temperate zones, the propensity of raised beds to absorb heat extends the growing season at both ends. Raised beds are also preferable from an ergonomic perspective in that they do not require the severe bending that a ground-level plot might. See chapter 3 for more details.

Square-foot gardening (SFG) is a subset of raised-bed gardening promulgated by Mel Bartholomew through a best-selling book of that name and popularized subsequently through television and other media. Rather than limiting one's effort to a square foot, what SFG does is divide a larger plot (usually 4 by 4 feet [1.25 by 1.25 m]) into individual square units. Planting in these units is dictated by plant size: for example, a unit might hold one tomato vine, or four thyme plants, or nine leeks, or sixteen carrots. As with many other (sustainable) intensive gardening methods, close planting helps shade out weeds and conserve moisture.

Hydroponics is the cultivation of plants through a nutrient solution, often in a soilless medium such as rock wool or expanded clay pellets. If the ubiquity of DDT helped galvanize the environmental movement in the 1960s, there was something else in the air by the 1970s: a high-grade purple haze. Marijuana is now the most profitable crop grown in the United States, and it is intimately connected with hydroponics. Although soilless growing techniques had been known for millennia, in the late 1960s and early 1970s, the political environment, technical advances, and—ironically—the spraying of the herbicide Paraquat over marijuana crops in Mexico encouraged domestic marijuana production. Hydroponic cultivation has been esteemed since then for its secrecy, ability to be used anywhere, and potency. In fact, hydroponics has been so closely linked to marijuana that a 2007 law-enforcement operation in Atlanta was dubbed "Hooked on 'Ponics."

Hydroponic techniques have many other applications as well. Eurofresh Farms in Willcox, Arizona, for example, produces more than 200 million pounds (close to 91 million kg) of pesticide-free tomatoes each year in one of the world's largest greenhouse complexes. Hydroponic techniques help conserve water and minimize pests.

Hydroponics incorporates a couple of different forms. *Aeroponics* is hydroponics that maximizes aeration of the roots through the application of a nutrient solution to the bare roots. *Aquaponics* is the integration of aquaculture with hydroponics. Although aquaponics has ancient antecedents—notably in Asian rice-paddy culture and Mexican chinampas—modern aquaponics focuses on a recycling system by which the waste of cultivated seafood is carefully calibrated to fertilize hydroponically grown crops. Some of the most promising developments for urban farming involve aquaponic systems that provide both vegetables and fish year-round. (See chapter 9 for more information.)

Urban Farming and Sustainable Production

Urban farming often combines a variety of growing methods (such as raised-bed or container gardening), philosophies (such as permaculture or biodynamics), and purposes (such as food production and education). For example, Florida's Bonita Springs Middle Center for the Arts (see page 70) has created an edible schoolyard, combining hands-on labor with food education that will ultimately include curricular objectives. They call it "a living classroom." The school employs many organic-friendly techniques—such as weeding by hand and laying down newspaper mulch—but is not dogmatically opposed to conventional fertilizer or other nonorganic approaches. The many tree crops grown at the school bring a permacultural feel to the garden, and the school hopes one day to let the students sell their harvest at the local farmers' market. Their edible schoolyard is not this or that; it is many things to many people. That is a typical quality of urban farms.

Brenda Tate, the volunteer who brought the edible schoolyard at Bonita Springs Middle Center for the Arts to fruition.

Sustainable Consumption

In addition to the production side, of course, sustainable agriculture involves consumption. Several movements and practices stand at the forefront of sustainable consumption.

The "Slow Food" and "Eat Local" Movements

The *Slow Food* movement began in Italy in the late 1980s "to counteract fast food and fast life, the disappearance of local food traditions, and people's dwindling interest in the food they eat, where it comes from, how it tastes, and how our food choices affect the rest of the world." According to Slow Food International's website, the movement aims for what it calls "good, clean, and fair food"—respectively representing taste, environmental impact, and compensation to producers. The movement has more than 1,000 local chapters all over the world,

A School Blossoms in Bonita Springs

Some people miss the forest for the trees. Some miss the trees for the forest. Brenda Tate sees both: the big picture and the little picture. Better still, she possesses an ability to join these visions to find a path to the other side.

When she moved to Bonita Springs, Florida, in 2006, the native Atlantan began mentoring local youth and raising funds for student scholarships. She knew how effective this combination was for kids living below the poverty line, and she saw things get worse when the recession struck. Many students experienced food insecurity. Instead of discouragement, Tate found inspiration.

Already familiar with Alice Waters's Edible Schoolyard project in California, Tate realized that in her area, the school year matches the growing season. She soon discovered a nonprofit called Florida Agriculture in the Classroom, which offers teacher training and curricular materials correlating with the Sunshine State Standards.

Most auspiciously, Tate knew that some 40 miles (65 km) north on I-75 lay the international headquarters of Educational Concerns for Hunger Organization (ECHO). Perhaps little known in the United States, ECHO supports poor farmers working in harsh conditions around the world through networking, technical assistance, a seed bank, and development of cheap but effective farming methods. Though the beneficiaries of ECHO's work are usually outside the United States, Tate convinced them to work with Bonita Springs Middle Center for the Arts, then known as Bonita Springs Middle School.

"I pledged from the get-go," says Tate, "that if they agreed to do this work I would make sure everyone knew of their mission and we would not become a distraction to that global mission." Not only did Tate approach ECHO but she also collaborated with a grantwriter friend to land a $3,000 grant from the Bonita Springs Community Foundation to make the project a reality. Local businesses and volunteers pitched in, too. Tate credits her background in project management as something that made it "very natural for me to bring all the pieces and all the players together." But the most dedicated workers were the students themselves. About two dozen students volunteered to work on the garden three days a week over the summer, notes vice principal Bob Scallan—and during a Florida summer, no less. Less than six weeks after breaking ground, students reaped the first reward of their labor: star fruit (*Averrhoa carambola*).

During the process of planning the garden, the school discovered that part of its campus was once a wetland. Like the garden itself, "The lessons to be learned from the project are growing and growing," according to Scallan. In fact, they have appeared to spread across Lee County School District. In 2016, for example, the Florida Commissioner of Agriculture recognized Trafalgar Middle School in its Golden Shovel Awards. The school distributes fresh produce to hundreds of families in need and won the "Best Use of Produce" award for secondary schools.

including more than 150 chapters and 6,000 members in the United States. While not specifically advocating urban farming, the movement's disdain for industrial agriculture and preference for local foods make urban farming a natural fit. It has also influenced other movements celebrating seasonal—and especially local—food.

Urban farming is, perhaps, most closely aligned to the Eat Local movement. In fact, "a viable and fairly common definition of urban agriculture," writes Jac Smit, who is considered by some to be the father of modern urban agriculture, "is that which is within 'same day delivery'"—in other words: local. Though eating local food was the default for most of human history, local food as a philosophy or movement is a much more recent development. It is a coalescence of many influences, including the Slow Food movement, with its celebration of local food traditions, environmental concerns, a reaction to the continuously dwindling number of small family farms, and the explosion of community-supported agriculture (CSA) programs and farmers' markets (which could equally be described as reactions to the local food movement). The underlying theme of local food was popularized by several books recounting authors' attempts to eat only food produced locally, such as Gary Nabhan's *Coming Home to Eat* (2001); Barbara Kingsolver's *Animal, Vegetable, Miracle: A Year of Food Life* (2007); and *The 100- Mile Diet: A Year of Local Eating* (2007) by Alisa Smith and J. B. MacKinnon.

Discussions about eating locally frequently use the concept of *food miles*, or how far food travels before it reaches its ultimate consumers. Though hard to quantify accurately, it is generally thought that the average item of food travels more than 1,500 miles (2,414 km) before it enters your mouth—meaning that, in all likelihood, your meal is better traveled than you are. The big issue in food-mile analysis is not the distance itself, but the amount of energy (and resulting greenhouse gases) needed to move that food. Professor David Pimentel of Cornell University has calculated that every calorie of food consumed requires several times that number of calories in production and transport—some much more than others—and that the average American's diet requires the equivalent of about 500 gallons (1,892 liters) of oil per year to produce.

Chef, author, and teacher Jessica Prentice contributed to the movement immeasurably by giving its practitioners something to call themselves: *locavores*. Locavores are not, however, some homogenous group. For one thing, the definition of "local" varies significantly: Nabhan experimented with 250 miles (402 km), for example, while Café 150 at Google's headquarters uses (you guessed it) 150 miles (241.5 km). Prentice's original Locavores group set the range at 100 miles (161 km); many other local-food enthusiasts aim for 50 miles (80.5 km); and Kingsolver limited her experiment to her own yard and nearby farms. Locavores also vary in their diets (omnivore, vegetarian, vegan) and in the degree of strictness with which they adhere to in eating locally.

Prentice herself is anything but doctrinaire. "I don't use the word as a whip to make myself or anyone else feel guilty for drinking coffee, cooking with coconut milk, or indulging in a piece of chocolate," she wrote for *Oxford American Dictionary* in relation to its selection of *locavore* as the word of the year in 2007. "There are things it makes sense to import because we can't grow them here, and they're either good for us or really delicious or both. But it doesn't make sense to watch local apple orchards go out of business while our stores are filled with imported mealy apples."

Farm-to-Consumer Approaches

As Dickson Despommier put it, "farming sucks." It is a tough business, particularly for small producers in the age of international agribusiness. Of all the money we spend on food grown in the United States, only about 20 percent goes to

Farmers markets have grown increasingly popular in cities.

the farmers. Two of the best competition strategies for the growing number of small farmers are to produce premium (biodynamic or organic), local, high-value food—such as gourmet greens and other perishable produce—and to build close relationships with consumers. Urban consumers can connect with urban and periurban farmers and support local agriculture by participating in community-supported agriculture programs, patronizing farmers' markets, and joining food cooperatives. All help small-scale farmers while providing urban consumers with a means to buy local food, even if they're not growing it themselves.

Community-supported agriculture (CSA) evolved in Japan and Europe in the 1960s in response to the continued industrialization of agriculture and its removal farther and farther from the consumer. Called different things in different countries, the basic concept of CSA remains the same the world over: consumers buy a "share" of a particular farm in advance of the growing season, guaranteeing them a proportional share of the harvest. It's good for the farmer, who is paid in advance, because the risk of misfortune is spread among shareholders. If a drought, a flood, or pestilence destroys part of the harvest, shareholders receive less than anticipated, but the farmer is still paid.

On the other hand, if there is a huge surplus, shareholders get paid back handsomely. CSA gives consumers security in knowing where their food is coming from and maybe even in knowing who is growing it. Some CSA shareholders volunteer to work on the farm or do so as part of purchasing a share. It is a popular way to reconnect with the land, with the seasons and their foods, and with other people.

Millions of Japanese and Europeans now belong to CSAs, but the concept has only taken flight in the United States since 1984. USDA data indicates that there are more than 12,500 farms employing some form of CSA.

Farmers' markets are another venue for farmers to sell directly to consumers, and one of the most promising avenues for the small, independent producer. According to the USDA, the number of farmers' markets almost tripled from 1,755 in 1994 to 4,685 in 2008. The number of farms actually increased over the same period by 1 percent, despite a declining trend since the end of World War II, and the average farm size decreased from 440 acres (178 hectares) to 418 acres (169 hectares).

Food cooperatives, or *co-ops*, are member-owned groceries. Unlike national or international supermarkets, they are often focused on their communities and dedicated to sustainable (often local) agriculture. Local Harvest's website (www.localharvest.org) has an extensive database of local CSAs, farmers' markets, farmstands, and co-ops nationwide.

How to Prioritize?

The original group of local-food enthusiasts for which Jessica Prentice coined the term *locavore* offers a hierarchy for choosing among foods at their website, www.locavores.com.

Locally produced is rated best, followed by organic (if you can't find locally produced), then family farm (or co-op of small farmers), then local business (especially for things that might not be economical to produce locally, such as coffee in most parts of the country), and finally "terroir." A term most familiarly used for wine, *terroir* describes the reflection of a certain place's uniqueness in the food it produces, such as sparkling wine from Champagne, coffee from Kona, or onions from Vidalia, Georgia. The point is not luxury but support for a local food—even if it's not your locality.

Similar to a small CSA or a food co-op but without the building, *buying clubs* pool member resources to buy in bulk the kind of food they want, such as local or organic. The goods are then dropped off periodically (such as once a week or month) at a location central to the club's members, who are usually responsible for distributing it among themselves.

Warning: Labels

Because so many of the philosophies and methods of sustainable agriculture react to the agricultural (and sometimes economic) status quo, it is tempting for those outside the sustainable-agriculture crowd to lump their proponents into a single category and ignore the nuances. In addition, keeping track of everything is confusing: there are methods (raised-bed gardening or hydroponics); movements, both loosely knit (Eat Local) and tightly knit (Slow Food); philosophies wed to certain techniques (biodynamics); whole disciplines (agroforestry); for-profits; and nonprofits. In contrast, those within the community are tempted to class everyone into tightly circumscribed boxes, with no room for variation. Paul Roberts, author of *The End of Food* (2009), has even called "[t]his tendency to replace complexity with checklists...[the] hallmark of the alternative food sector."

Issues of food, agriculture, sustainability, and globalization occupy the minds of many passionate people. They may be on a pilgrimage to the same place—sustainability—but they come from different places and move at different paces, and sometimes they step on each other's toes along the way. As much as sustainable agriculture's proponents celebrate diversity, the achievement of real sustainability necessitates trade-offs and compromises that will leave some disappointed.

For example, can hydroponics truly be organic? It certainly can be implemented without prohibited inputs. But many traditional farmers and organic enthusiasts see soilless agriculture as at odds with the spirit of organic agriculture, the very name of which derives from viewing the soil as an organism. Only in 2017 did the National Organic Standards Board make the call, finally deciding that, yes, hydroponic and other out-of-soil crops can be designated organic.

Many practitioners of the philosophies-cum-methods previously mentioned, such as biodynamics, traditional organic agriculture, and natural farming, find in the ground what others find in the stars: something beautiful, unthinkably complex,

Hydroponics allows the controlled production of plants in virtually sterile environments.

and almost magical. In a section of her book *Food Not Lawns* (2006) called "Dirt Worship," H.C. Flores writes, "We can begin to heal our relationship with nature when we overcome our fear of the dirt."

However, many people approaching urban farming from a more clinically logistical standpoint—feeding people safely and sustainably—favor contained, soilless techniques. Despommier's vertical farm concepts, for example, are completely hydroponic. This is not out of any lack of appreciation for the soil's life—in fact, quite the opposite. Despommier is a parasitologist who has long sought to help the tropical poor, hundreds of millions of whom are plagued by geohelminths, parasitic worms spread primarily through traditional agricultural activities.

Tilling versus no-till agriculture presents another point of debate. Except for an initial investment in specialized seed-drilling equipment, no-till agriculture seems at first almost like a panacea for the ills of modern crop management: it reduces time spent in the fields, oil and water consumption, soil erosion, the need for fertilizer, and pollution through runoff. In this way, it is very much in line with ecologically minded practices such as biodynamics and organic agriculture, yet certification in either is very challenging for a no-till farmer because of weeds—or, more specifically, their elimination. Tilling reduces weed populations significantly, while no-till methods let them run rampant. As a result, no-till farming often calls for relatively higher use of herbicides, which suddenly makes crops genetically engineered (GE) to resist herbicides all the more attractive, since one can spray the whole field and not worry about targeting only weeds.

Specific technologies such as hydroponics, no-till farming, and GE crops may divide the sustainable-food movement into different crowds, but so may technology itself. David Holmgren, one of permaculture's founders, writes that "[m]uch of the optimism about sustainability relates to the application of technology and innovation. Permaculture strategies make use of these opportunities while maintaining a healthy skepticism based on the premise that technological innovation is often a 'Trojan horse,' recreating the problem in new forms." Think of the hybridization of European and African honeybees to develop productive tropical honeybees, which instead became "killer bees." Or consider the release of cane toads to control cane beetles in Australia, which had no effect on the beetles but instead littered the continent with quickly reproducing, poisonous toads. You get the idea.

Slow Food takes a similar stance on the organic movement. Although it favors sustainable practices such as those common to organic farming, according to the Slow Food Gold and Treasure Coast website (www.slowfoodgtc.org), it "maintains that organic agriculture, when practiced on a massive and extensive scale, is very similar to conventional monoculture cropping, and therefore organic certification alone should not be considered a sure sign that a product is grown sustainably."

No-till soybeans in a field in Argentina.

If you've ever tried to grow food organically, you know why organic produce costs more.

Slow Food's concerns may be borne out. Organic production has expanded dramatically, but largely thanks to the same agribusinesses that control conventional agriculture. As sustainable-food author and consultant Cynthia Barstow observes, looking back on her book *The Eco-Foods Guides: What's Good for the Earth is Good for You!* (2002) and concomitant tour, "I now know that my little effort was a blip on the screen compared to what one multinational can do with one little can of tomatoes labeled with the organic seal."

This success strikes many as a double-edged sword. On the one hand, it means that millions of additional acres now grow food without the synthetic fertilizers and pesticides to which sustainable agriculture advocates have long objected. On the other hand, the vision of agribusiness being beaten down by a wave of small organic family farms has not materialized. In fact, agribusiness often advocates for stricter standards, according to Organic Trade Association founding member Grace Gershuny, who also worked with the National Organic Program. "Big companies have compliance staffs to make sure they are meeting all the requirements of myriad types

of certifications," she notes. In her experience, in fact, it's been "usually the small, independent folks who felt they needed the rules bent to accommodate their needs." The more hurdles, the greater the advantage to big players.

For some, our economic system itself is the enemy. Jessica Prentice told me about this kind of conflict in connection with opening Three Stone Hearth, a community-supported kitchen in Berkeley, California. Though the mission of Three Stone Hearth is nonprofit in spirit, it is legally for-profit—technically a worker-owned cooperative—because our economic system favors the development of businesses. (Another example is Garden State Urban Farms in Newark, New Jersey, also deeply invested in the community, which opted to be for-profit for similar reasons.) Three Stone Hearth's mission doesn't even mention profit (Prentice says she isn't in it to retire wealthy), yet even this unorthodox model—clearly devoted to the community and environment—raised eyebrows among some community members, as if Prentice and her partners were enemy collaborators for working with "the system."

People choose to be locavores for many reasons—among them the taste of local meats and produce, support for community, and dislike of industrial agriculture—but one of the most common is environmental: diminishing one's ecological footprint. Yet a recent study found that food choice determines one's footprint more than food miles do. It specifically identified meat and dairy as the real greenhouse culprits—the bigger the source, the worse, with the cow being something close to a dietary Voldemort. The study found, for example, that shifting your red meat and dairy calories to chicken, eggs, fish, or vegetables just one day a week reduced greenhouse gases far more than a 24/7, all-local diet. How should an environmentally conscious, omnivorous locavore respond?

In food as in life, people generally don't fit neatly within labels and all they imply. I recall a relative who worked in the food industry (but outside sustainable agriculture) telling me that he knew for a fact that organic food was no fresher or better tasting than conventional food. Given that he knew my wife and I had bought organic produce at Whole Foods, this was sort of a declaration: *Aha! All you believe in is false!*

The statement involved a cascade of assumptions: (1) because we had shopped at Whole Foods, we must do so all the time (false); (2) all we would buy there would be organic (false—who could afford it?); (3) our primary reasons for buying organic produce were freshness and taste (false); and (4) we probably did things like drink raw milk and wear organic hemp clothing (false on both counts). Yet, strangely enough, it had never occurred to me that people might shop for organic produce *primarily* for freshness and taste. We had an infant at the time, and our bigger concern was consuming fewer pesticides in produce typically grown with lots of them. But we didn't buy exclusively organic, and we would almost certainly choose fresh, conventionally grown, local food over old, organic, imported food.

One of my best friends, incidentally, does favor raw milk and organic hemp clothing. To many, this would conjure up the image of a crunchy lacto-vegetarian; in fact, he not only eats meat, he often eats *raw* meat, and he considers all types of vegetarians to be seriously jeopardizing their health. And the same relative who challenged me with the foregoing organic-produce comment later opened a vegan ice cream shop. In researching this book, I found a new agrarian who cited biodynamic farmers as among the "weird elements" of his movement, an agroforestry expert who found some permaculturists to be "polluting" gardening and the wider culture, and an organic landscape designer who was opposed to recycling (and, to be fair, using) plastic. The sustainable-agriculture crowd can make strange bedfellows.

As a pillar of sustainable agriculture, urban farming provides a common ground to advance the goals of many disparate groups, from rural agrarians to new urbanists, from neighborhood activists to city-hall planners, from food-justice liberals to national-security conservatives, and from small independent farmers to international venture capitalists. The biggest question may be how urban farming will affect its rural counterpart.

An End to Rural Agriculture?

Urban farming poses little threat to rural agriculture. The two are, in fact, rather complementary. Most experts see urban farming as a means not to total sufficiency in food but to self-reliance—being able to produce enough food to contribute meaningfully to trade or local consumption in good times, and to provide some fallback food security in challenging times. Cities can reasonably produce many perishable fruits and vegetables, eggs and chickens, possibly fish and milk, and a range of smaller livestock and other goods suited to their particular tastes and infrastructure. Yet rural areas will still dominate in staple grains, large meat livestock, nonperishable foods, and crops such as sugar cane that have restrictive growing requirements. Urban farming will not necessarily make rural agriculture sustainable, but it can bring significant advantages to rural communities.

Perhaps the most obvious advantage is the dramatic reduction, facilitated by urban farming, in transportation-based air pollution, which benefits everyone. In addition, rural lands and aquifers suffer most directly from industrial agriculture; urban food production can lessen some of the destructive pressure to squeeze the last bit out of every patch of rural earth. Cities are better positioned to reuse water resources, potentially saving aquifers and reservoirs in rural communities. Finally, dense urban life tends to be much less energy-consumptive per capita than rural life. "Living surrounded by concrete is actually pretty green," notes economist Edward Glaeser. So every person needing a home who is accommodated in the city rather than in the country helps preserve the natural resources of rural communities while simultaneously lowering the consumption and pollution that impact us all, wherever we live.

As urbanization intensifies, small rural and suburban farmers can benefit from farmers' markets, CSAs, buying clubs, and co-ops near city fringes. The concentration of people in cities also brings new opportunities for rural agritourism. Roughly 2,000 agritourism sites in and around Beijing draw tens of millions of visitors each year. Could genuinely rural farms be any less attractive to urban dwellers over here? Homegrown agritourism would also tend to favor smaller producers because nobody wants to visit a massive feedlot.

Urban farming makes cities stronger and more stable, with higher employment rates, better nutrition, less pollution, and enhanced resilience to economic, political, and meteorological blows. This translates to better markets and a less-dependent nation. It would be absurd to say that urban farming is the savior of rural agriculture with all of its problems, but it is certainly not its enemy. In city or country, for omnivore or vegan, business-as-usual agribusiness is simply unsustainable. The first and best step toward sustainability may be the urban farm.

CHAPTER 3

Toward an Urban Farming Future

Before considering what the future of urban agriculture might resemble in the world's richest country, it's instructive to consider the absence of urban agriculture in one of the world's poorest countries: Haiti. On January 12, 2010, a magnitude 7.0 earthquake struck Haiti, devastating Port-au-Prince, Jacmel, and other cities. An estimated 230,000 people died. That's a breathtakingly high number. Stronger earthquakes have hit more densely populated cities with a fraction of the casualties. Fewer than 1,000 people died in the magnitude 8.8 earthquake in Chile the following month, for example, and that was about 500 times as powerful.

This wild discrepancy involves many factors—planning, geography, infrastructure, international trade, and plain old luck, to name a few—but most involve a vast difference in wealth. Related

to these factors but worth exploring itself is the story of agriculture. Haiti has very little. Would urban farming have significantly cut the death toll? Absolutely not. Physical trauma was almost certainly the leading cause of death, with infection a distant second. No, the lack of agriculture did not contribute to the immediate damage of the earthquake, but what happened in the earthquake's aftermath speaks volumes about the lack of agriculture.

Despite the sometimes-miraculous recoveries of trapped individuals, the most gripping aspect of the earthquake quickly became the care of the hundreds of thousands of survivors. And with a spell of forgiving weather postquake, food and water surpassed shelter as the most pressing necessity. Who would have imagined that it would take so long to deliver food and—on an island plagued by flooding—water? Yet food has been in short supply in Haiti for decades; it was one of the countries beset by riots over food prices in 2008. The country used to grow enough to support itself but now imports about three-quarters of its food. Although the hilly terrain makes tree crops especially attractive, the country is 98 percent deforested due to historical exploitation of timber resources, the ongoing demand for charcoal, and poor stewardship. The deforestation, in turn, feeds the water problem. When it rains, it floods, causing mudslides and further erosion of cropland. Deforestation also removes a biological filter that would help provide a clean water supply. It's a downward spiral.

One lesson to draw from Haiti's experience is the importance of a diverse, distributed, well-managed food supply. In other words, *sustainable*: urban and rural, annual and perennial, large-scale and small-scale. Almost two-thirds of Haiti's population comprises farmers. That would make it seem as if Haitians are very agricultural, but numbers are deceiving: the farmers are mainly subsistence-level—they grow what they can to eat and because there aren't many other avenues for employment.

Subsistence is just about the one reason that does not motivate urban farmers in the United States, but it's worth remembering that whatever their individual reasons for farming, food is a by-product, and the accumulation of these by-products contributes to our food security. Could you wait two weeks for an airdrop of food if things hit the fan in your city? That's a good reason to advocate for sustainable agriculture in our country.

Well, things aren't going to "hit the fan" here, you might think. And you're probably right. But another lesson from Haiti is just how precarious and close to us food insecurity is. In fact, the earthquake in Haiti seemed very much like the fulfillment of the scenario from chapter 1: an unexpected blow hitting an unprepared government in a nation already ineffective at feeding, protecting, and providing basic services to its citizens. Then, as predicted, nonstate groups, such as gangs, gain power, and the next stop is a failed state just 700 miles (1216.5 km) from Miami. Assuming George Washington was correct in observing that nations don't do each other favors, then fear of a failed Haiti seems a good explanation of why the United States sent more than 15,000 troops there at a time when they were already engaged in two wars. That's a good reason to advocate for sustainable agriculture in other countries.

Many plans for the rebuilding of Haiti called for agriculture—the commercial-scale, well-managed kind—as a way to generate real jobs, increase food independence (through local food staples), and generate income (through crops such as tropical fruit, which can be sold abroad). It was, in effect, a return to an agricultural tradition but with modern sensibilities, an embrace of the notion that progress need not conflict with tradition. In other words, it was similar to what was (and is) happening across the United States at different stages in different cities.

Celebrity chef Rick Bayless (see chapter 8) recalled being at a restaurant in San Juan, Puerto Rico, that proudly heralded the source of its ingredients—ingredients from everywhere, it seemed, but Puerto Rico. He learned that there was little agriculture left on Puerto Rico and that even much of its famous rum was made from imported sugar. It struck him as so different from Chicago, where local food was all the rage among chefs. He likened the difference between Puerto Rico and Chicago to traveling back more than three decades in time. He recalled, "What people are all talking about is 'Oh, well, we only serve New Zealand venison,' and '[We] fly in our goods from here,' ... and it's all from somewhere else." The key difference is that Chicago returned to its rich agricultural tradition (after diverging from it just decades ago in favor of imports) not by necessity, but by choice. One challenge of urban food systems today is to restore the capacity of (and pride in) local agriculture while more responsibly importing the diversity of products that can come only from elsewhere. It is an approach that is less "either/or" and more "both/and."

The State of Urban Farming in the United States

A central observation of this book is the sheer diversity in farms, farmers, reasons for farming, and ways of doing so. Although the boom in urban agriculture is impossible to document with precision, the circumstantial evidence is impressive, as highlighted in the introduction to this updated edition. In addition, thousands of urban schools participate in Farm to School, Edible Schoolyard, and similar projects. Consumers have purchased millions of EarthBoxes (specially designed troughs used for container gardening). Chicago alone has more than 500 vegetated rooftops with nearly 6 million square feet (557,400 sq m) of green roofs, some of it, like Rick Bayless's rooftop gardens, devoted to food.

We know from USDA figures that the number of small farms is increasing nationwide, but it's hard to know how many of these are urban and whether or not most urban farms are included in the data; I'd guess not. We can try to extrapolate from known figures. The American Community Gardening Association's (ACGA) database, for example, lists fifteen community gardens within 10 miles (16 km) of Detroit. Meanwhile, the Garden Resource Program Collaborative in that city cites that there are more than "1,500 productive urban gardens and farms," which is 100 times as many as the ACGA lists. If this underrepresentation holds true for other cities, then the ACGA's figure of 18,000 community gardens suggests that there might be closer to two million community gardens and farms across the United States and Canada.

Data hints at the strength of urban farming in the United States, but there's no central authority, no single source of information. There's no pope of urban agriculture. One block's farmer may not even know what's happening on the next block. This was vividly illustrated at one of the first sites I visited as I was writing

When you think of Jersey City, New Jersey, you probably don't think of the delicious-looking vegetables at this farmer's market.

this book. It was a small plot in the South Bronx at a residential home run by The Bridge for people with mental illness, some of whom farmed the site. I asked the site coordinator if she knew of any other local farms, and she said she'd heard of another in the same neighborhood but couldn't say where. She then asked one of the farmers, who pointed to a fenced, wooded lot less than 200 yards (183 m) away. More recently, the headquarters of AeroFarms—and site of one of the world's largest indoor vertical farms—opened at the location of a former steel mill in Newark, New Jersey, mere miles from my home.

Farms can be within a stone's throw of each other and not even be aware of each other. This makes urban farming very lively, and—at the risk of being cute—organic, but it also limits impact because there is no "face" to push policy changes to make it more effective. Things are changing, however. Cities are jockeying to become leaders in the field, foundations are funding urban farming in the hopes of solving a panoply of city woes, and new hands are putting on garden gloves every day. For all its quirks and challenges, urban agriculture definitely seems on the way to "conceptual maturity," as evidenced by its growing prominence at city halls, among academics, and in the business world, as well as by the very visible success of some larger farms.

City Halls

While most cities are probably taking a "wait-and-see" approach or planting a token garden at city hall, others are already positioning themselves as the future leaders in urban agriculture. One example is Seattle, which, culminating several years of innovative efforts, declared 2010 "The Year of Urban Agriculture" and launched an informative website (www.urbanfarmhub.org) to detail its programs. The city provides gardens that residents can farm, for example, and a CSA through which they can sell their locally grown food. Seattle also allows gardening on the strips between the road and one's property and offers neighborhood groups the opportunity to apply for free trees, including fruit trees. The city's 2017 comprehensive plan mandated at least one community garden for every 2,500 households in a given area.

Seattle is not alone. In big ways and small, cities such as Portland, Oregon; Boston; Milwaukee; Madison, Wisconsin; Kansas City; Chicago; Baltimore; and many more have taken steps to make urban agriculture more accessible. New York City, no laggard itself, also shows how urban agriculture leaders encourage other cities to follow suit. A report—"FoodNYC: A Blueprint for a Sustainable Food System"—released in 2010 by the office of the Manhattan borough president recommends, among other things, that city agencies annually assess city property suitable for urban agriculture (such as in Portland), develop a municipal urban agriculture program (such as in London), permanently preserve community gardens as parks (such as in Cleveland, Boston, and Seattle), and facilitate rooftop greenhouses (although the report doesn't acknowledge Chicago in this aspect, it is the green-roof king).

Wider Expert Acceptance

City government is complex and often overlapping. For example, should urban agriculture fall within the city planning department, or parks and recreation, or zoning, or perhaps neighborhood and community services? It probably falls under all of them in some capacity or other, and each of these departments has its own teams of professionals. While urban agriculture might once have been the unchallenged domain of hippies, foodies, guerilla gardeners, and miscellaneous inner-city do-gooders, it is quickly being integrated into the disciplines responsible for running cities.

A bit of a watershed year was 2009, with urban agriculture figuring into the conferences of the American Planning Association in Minneapolis, the Urban Affairs Association in Chicago, the Third International Conference on Healthy Food in Healthcare in Detroit, and the seventy-seventh annual meeting of the US Conference of Mayors in Providence (where they adopted not one but two resolutions in favor of urban farming). And that doesn't even include many citywide and regional conferences, such as those in West Valley City, Utah; New York; Philadelphia; and many other places. Since then, conferences have only increased in number as well as in specialization for farms and farmers. For example, Newbean Capital has run its biennial Indoor Ag-Con since 2013, and Black Urban Growers (BUGs) has hosted an annual Black Farmers and Urban Gardeners Conference since 2010.

The number of books, websites, blogs, podcasts, and other resources related to urban farming have skyrocketed. Theses and dissertations on urban agricultural issues have proliferated. At least a dozen institutions of higher education now offer degree or certificate programs in urban agriculture, and other universities are sure to follow suit. As I was writing the first edition of this book, Martin Bailkey, a food systems planner and coauthor of *Farming Inside Cities: Entrepreneurial Urban Agriculture* in the United States, told me that while keeping up with all of the happenings in urban agriculture used to be an extracurricular pursuit, it would now "take someone working full time with support. Now it would take a team, and a large one at that."

Related Commercial Ventures

Another very promising sign has been commercial enterprises geared primarily toward urban food production. Intricately connected with this planning is the recognition that rural croplands are exploited without rest, while city spaces remain perversely fallow. This is beginning to change, and many see the new horizon as vertical.

Some visions of these urban agricultural efforts—what writer Bina Venkataraman has accurately described as "hybrids of the Hanging Gardens of Babylon and Biosphere 2 [one of the world's largest enclosed ecosystems] with SimCity appeal"—look futuristic, which a pessimist might define as "never going to happen." Yet human space travel looked futuristic in the 1950s—not because the achievement was impossible, but because it was untried.

The decades since then have seen huge technological strides that would enable these projects, including those in the areas of tissue micropropagation and genetics, hydroponics, LED lighting, biofuel and other chemical processes, the continued evolution of computers, and supermaterials such as ethylene tetrafluoroethylene (ETFE). EFTE is a polymer that is extremely strong, clearer and 99 percent lighter than glass, able to stretch three times its length while maintaining elasticity, and even recyclable. Inflated pillows of ETFE cloak the spectacular Water Cube (pictured) from the 2008 Beijing Olympics as well as the Eden Project facility in the United Kingdom.

Sky Vegetables is a company with a vision both ambitious and eminently practical: growing fresh produce locally in rooftop greenhouses. After developing its nutrient film technique (NFT) hydroponic growing method in Massachusetts, in 2013 the company opened its first commercial farm atop a Leadership in Energy and Environmental Design (LEED)-certified housing complex in the Bronx. The company produces about a ton of produce a week from its 8,000-square-foot (743-sq-m) greenhouse, using recycled rainwater, and "wants to scale their building-integrated agriculture concept to other rooftop farms in New York City and beyond."

Beijing's Water Cube

Imagine ever-moving waves of plastic lunch trays, and you'll envision something very much like the VertiCrop High Density Vertical Growth System, developed by Valcent Products in El Paso, Texas. Originally conceived as a way to save energy in greenhouses by using all three dimensions of the structure, instead of heating all three but growing in just two, the VertiCrop system moves plant trays through a rotating conveyor system that precisely calibrates the light needed for each crop. Water and nutrients are similarly parceled out, resulting in about twenty times the crop yield per acre as conventional agricultural fields while consuming just one-twentieth of the water. The fully contained hydroponic system is free of herbicides and pesticides, relying instead on sophisticated filters and ultraviolet lights.

Nearly 7,800 miles (12,553 km) away in Nimbin, New South Wales, Urban Ecological Systems (UES) has worked on another self-contained system, this one based on combining horticulture and aquaculture. As with aquaponic approaches long practiced in Asia, fish waste is used to fertilize plants. Unlike many traditional aquaponic systems, however, the one developed by UES discharges

no polluted effluent. The plants "strip" the nutrient-rich waste from the water, which is then returned, newly clean, to the fish. It produces several times as much yield as traditional cropland and, as with the VertiCrop system, it takes just 5 percent of the normal water use.

Growers can also now invest in modular, expandable turnkey systems, from a traditional-style greenhouse with cutting-edge nutrient film technique (NFT) hydroponics to an almost fully automated shipping-container-based system (standard-size shipping containers are modified to be little crop factories) using raft-solution hydroponics that can pump out more than 100,000 heads of lettuce a year. These kinds of systems may not be economically competitive in the short term, but as the prices of oil and clean water rise, they will most likely become so. They may also work well in circumstances in which a store or restaurant wants to grow its own food, with the premium on sales outweighing the premium on costs.

There are commercial opportunities at every stage of urban farm development. In the Greenpoint neighborhood of Brooklyn, for example, green-roof consultancy Goode Green installed a 6,000-square-foot (557.5-sq-m) rooftop farm, while another company, Gotham Greens, boasts its own 16,000-square-foot (1,486.5-sq-m) hydroponic greenhouse.

One of the most successful ventures may also be one of the simplest: specially designed plastic containers such as the EarthBox. These containers can easily grow most vegetables, guarantee clean soil (as long as you put in clean soil), and make water management a snap. Best of all, they can move. With land tenure being one of the biggest challenges for the growth of urban agriculture, that's a huge advantage.

Success and Proliferation of Large Farms

Another sign of the rise of urban agriculture has been the success of relatively large entrepreneurial farms and collections of farms. Of course, "large" means different things to urban and rural areas. The nonprofit Added Value runs two farms totaling 5.75 acres (2.3 hectares) in New York City, which is huge for such a densely populated city. Growing Power has a 2-acre (0.8-hectare) property in Milwaukee that is jam-packed with urban agricultural activities, another 5-acre (2-hectare) property in the city for a school and community partnership project,

Vertical doesn't necessarily mean multi-story; stackable pots such as these strawberry planters allow intensive growth in small spaces.

and an additional 40-acre suburban (16-hectare) property. It also runs three smaller farm projects in Chicago. Rio Grande Community Farm boasts an incredible 50 acres (20 hectares) within the city of Albuquerque. Most of these farms serve as venues for community-directed services such as job training, youth development, and educational programs in addition to providing food.

Even as large urban farms prosper in some cities, others host a vibrant group of smaller farms, thanks to partners such as the Kansas City Center for Urban Agriculture, the Garden Resource Program Collaborative in Detroit, or the New Orleans Food & Farm Network. Perhaps the biggest test, however, will be the success of for-profit urban growers, such as Gotham Greens or, on a much larger scale, Hantz Farms (see From Motown to Growtown sidebar in this chapter).

What Is Possible?

One of first questions of urban farming skeptics is just how much production is realistically possible in and around cities in the United States? And if the answer is that it will pale in comparison to rural agriculture—which it surely will—then the whole endeavor can seem like folly, and folly so passionately advocated smacks of a political agenda, and a political agenda suggests...well, I'll just let Joel Kotkin explain it. In an article called "America's Agricultural Angst," Kotkin, a distinguished presidential fellow in urban futures at Chapman University, writes:

> **...farmers have found themselves in the crosshairs of urban aesthetes and green activists who hope to impose their own Utopian vision of agriculture. This vision includes shutting down large-scale scientifically run farms and replacing them with small organic homesteads and urban gardens....**
>
> **A formula that works for high-end foodies of the Bay Area or Manhattan can't produce enough affordable food to feed the masses—whether in Minnesota or Mumbai. The emerging war on agriculture threatens not only the livelihoods of millions of American workers; it could undermine our ability to help feed the world.**

Central to Kotkin's thesis is the undeniable assertion that urban land is more expensive to buy than rural land. He also suggests that maintenance costs would probably be higher in cities. When you figure in savings on transportation and

Tomatoes thriving in
an EarthBox

refrigeration costs, exploitation of the heat island effect, and the potential recycling of a city's food and water waste—which could both cost less than virgin resources and avoid externalizing costs of disposal—it really depends; the relative maintenance costs could go either way, based on the price of oil and other factors.

Here, however, the conceptual mistake is in viewing urban agriculture as an alternative to rather than a complement of rural agriculture. One can be pro-urban agriculture without being anti-rural agriculture. Granted, many of the pro-urban agriculture crowd dislike agribusiness, which is primarily rural, but I believe most of them are looking more for an alternative—and a chance to get their hands dirty—rather than a revolution.

A small piece of urban land can accommodate a large number of plants.

Kotkin's more practical flaw is in discounting the urban space available for farming, especially when small, well-tended urban farmsteads can be so productive. On 1.5 acres (0.6 hectare) in Detroit, Earthworks produces about 3 tons of food and half a ton of honey annually. The roughly 4,400-square-foot (409-sq-m) garden (about 0.10 of an acre [0.04 hectare]) of the Dervaes family, famous urban homesteaders in Pasadena, produces about 3 tons of fruits and vegetables, 2,000 eggs, and 25 pounds (11.3 kg) of honey annually. In addition to their flagship 16,000-square-foot (1,486.5-sq-m) (0.45-acre [0.18-hectare]) hydroponic greenhouse in Brooklyn, Gotham Greens has built and operates three other urban rooftop greenhouses (another in Brooklyn, one in Queens, and one

From Motown to Growtown

The driver for American urban agriculture over the next decade may well be the Motor City. Detroit (pictured) has a thriving urban agricultural community, with as many as 1,500 community and backyard gardeners, according to Professor Mike Hamm, the Charles Stewart Mott chair for sustainable agriculture at Michigan State University. That would translate to almost one in 500 residents. (For comparison, the same ratio in Los Angeles would total about 7,200 gardens.) Detroit's farmers enjoy excellent technical support, thanks to the city's Garden Resource Program (GRP), which provides seeds and locally grown transplants, helps test soil, and offers other services to nearly 900 urban farms and gardens. The collaborative involves three major nonprofit groups, along with the Michigan State University Extension. Then there is the Detroit Black Community Food Security Network, which works with the GRP to develop an entrepreneurial model for 1- to 3-acre (0.4- to 1.2-hectare) farms, among other things, and Urban Farming, an international nonprofit based in Detroit that has close to 65,000 partner gardens as part of what they call their "global food chain." And that's not all.

It's not necessarily just the enthusiasm of farmers that makes Detroit a kind of urban-farming mecca. It is also desperate need—and the support of ways to address that need. With nearly 140 square miles (363 sq km) and a prime location on the Detroit River, which links the Great Lakes, Detroit was built as a city for about 2 million people. After peaking at about 1,850,000 in 1950, however—when it was more densely populated than contemporary Chicago—it began hemorrhaging residents. A report by the American Institute of Architects Sustainable Design Assessment Team ("Leaner, Greener Detroit") predicted that the population may ultimately stabilize at as low as 500,000 residents—a quarter of the residents for which it was built.

It is estimated that more than one-quarter of Detroit's residents live below the poverty line. It is also the paradigmatic "food desert." There are no major supermarkets in the city, and more than 92 percent of the stores that accept food stamps are "fringe retailers" such as liquor stores and gas stations. Worse still, the 8 percent of actual food stores average two or more times the distance away as fringe locations for the

majority of the city's population. As "Leaner, Greener Detroit" puts it, "From a symbol of industrial dynamism, Detroit has become a byword for economic decline and urban decay."

Although the unemployment rate has steadily improved since the the publication of this book's first edition, Detroit lacks money, fresh food, and a tax base. What it has in abundance is enthusiasm, a hunger to work, and land—more than 40 square miles (104 sq km) of vacant lots. At least 9 square miles (23 sq km) of that—about 5,500 acres (2,226 hectares)—comprises publicly owned but unused large plots, not even counting parkland, private land, or the land associated with the tens of thousands of abandoned buildings that litter the city. Professor Hamm and visiting specialist Kathryn Colasanti estimated that by working just 2,000 acres (809 hectares), Detroit could produce three-quarters of residents' vegetable needs and about half of their fruit needs. Working 10,000 acres (4,047 hectares), Detroit could—in the words of "Leaner, Greener Detroit"—"support

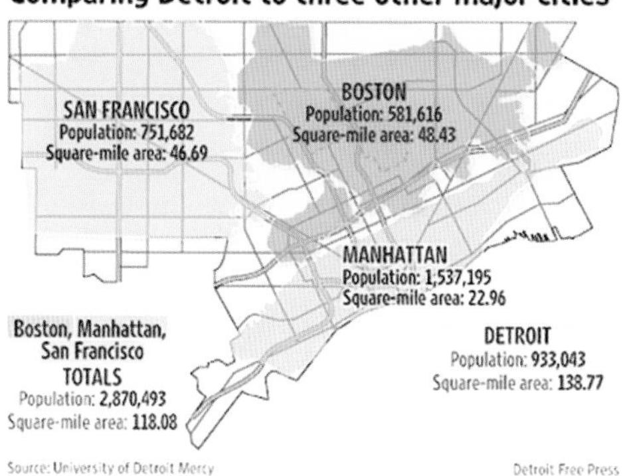

Professor Dan Pitera of the University of Detroit Mercy illustrates just how large and relatively unpopulated Detroit is compared to other major cities.

hundreds of farms and generate thousands of jobs, while dramatically improving the health of Detroit's residents.... Detroit should be able to build an urban-agriculture system that would substantially exceed any other system in the United States."

All of these factors make Detroit a perfect testing ground for agricultural models of varying motives, sizes, and ownership structures. The Majora Carter Group, for example, proposed the development of a worker-owned cooperative, as Ocean Spray is for cranberry growers, Land O' Lakes is for dairy farmers, and Blue Diamond Growers is for almonds. Coordinating hundreds or thousands of small farmers, it could make urban agriculture a truly profitable venture. A coalition led by the SHAR Foundation, which is affiliated with the nonprofit Self-Help Addiction Rehabilitation (SHAR), proposed farming up to 2,000 acres (809 hectares), starting with a 30-acre (12-hectare) pilot farm. The RecoveryPark project aims to ultimately hire thousands of workers and integrate housing development, other commercial enterprises, and nonprofit uses such as extended support services (including jobs) for recovering addicts. Its first business, RecoveryPark Farms, was established in 2012, and other farming projects have followed. The venture has successfully built relationships with many of the city's top restaurants, supplying them with fresh local produce.

The biggest and most contentious plan, however, has been a commercial venture by Detroit financier John Hantz, with plans to create nothing short of the world's largest urban farm—a kind of "agricultural Disney World," as Hantz Farms president Mike Score has put it. The venture acquired some 1,500 vacant lots—more than 140 acres (about 57 hectares) of land—in 2013. Since then, Hantz Farms has cleared scores of blighted properties and planted tens of thousands of hardwood saplings as part of a "Hantz Woodlands" initiative. Nevertheless, Hantz Farms—established by a rich, white man buying up a huge swath of a predominantly poor, black city at rock-bottom prices—remains controversial.

Whatever the motive for farming, Professor Hamm believes (without playing favorites) that Detroit would need a combination of large, medium, and small farms to reach a threshold of about 2,000 acres (809 hectares). "Given the economic situation in Michigan," Professor Hamm said in an interview with the C.S. Mott Foundation, "we don't have the luxury of pursuing a single goal in our work. As a state, we are being forced to achieve multiple goals through one activity—rebuilding our economy. We [the C.S. Mott Group for Sustainable Food Systems] believe agriculture, and, more specifically, locally integrated food systems, will be a part of that new economy."

If Detroit can make it happen, it just may achieve Urban Farming founder Taja Sevelle's vision of "turning Motown into Growtown."

in Chicago), totaling an impressive 170,000 square feet (15,793.5 sq m) across all four locations.

There is a surprising amount of room for urban farming in many cities. According to one survey, about 15 percent of land, on average, in US cities is vacant. Just how much and where it is depend largely upon the city and vary considerably. Consider the example of coastal cities flanked by mountains. The city and borough of Juneau, Alaska, encompasses more than 3,200 square miles 8,288 (sq km) and has a population density of fewer than fifteen people per square mile. If property were divided equally, every resident would have about 55 acres (22.25 hectares) of land and 10 acres (4 hectaares) of water. There's no need for space-saving vertical farming in Juneau, although hoop houses and greenhouses would be helpful climatically.

To compare, each of Hong Kong's 426 square miles (1,103 sq km) has more than 15,000 people (about 25 people for every acre or about 55 people for every hectare). Hong Kong is also significantly warmer, and a plot can reportedly grow six crops of cabbage in a given year. Metropolitan Vancouver is a blend: warmer than Juneau, but cooler than Hong Kong, with an intermediate population density of about 1,900 residents per square mile (2.6 sq km) (roughly 3 people per acre or 8 people per hectatre). It also happens to be one of the most advanced cities in the Western Hemisphere for urban agriculture.

Even in places without 2 acres (0.8 hectares) you could rub together, there is often vacant space under power lines, next to train tracks and roads, and in other out-of-the-way areas. The United Kingdom has 600 acres (243 hectares) of window boxes. New York City has about 14,000 acres (5,665 hectares) of unshaded rooftop and 52,000 acres (21,0433 hectares) of backyard; if all of that were used as productively as Earthworks's space, it would produce almost 135,000 tons of fruits and vegetables, or about 30 pounds (13.5 kg) of produce (and 5 pounds [2.25 kg] of honey) per city resident. And that's not counting the potential of street trees, empty lots, or parkland.

With an ancient and enduring tradition of urban agriculture, it should not be surprising that Chinese cities are leaders in the field. Much of this success is entirely modern, however, and relates to efforts to end a history of food shortages. The land area of Beijing exploded to more than 250 times its size in 1949, for example, as central authorities pushed out the boundaries of cities to bring permanent periurban agricultural areas under the administration of cities. As a result, many major Chinese cities have very dense urban cores surrounded by much less dense suburban, or even rural, areas. Urban agriculture is built in by design, or fiat, depending on how you look at it. It is an integral part of major Chinese cities.

In contrast, urban agriculture reportedly "never figures in the master plans of Indian cities; on the contrary, urban agriculture is viewed as backward and something to be minimized," according to Yue-man Yeung, a professor of geography at the Chinese University of Hong Kong. Similarly, many African cities, while often large and less densely populated, have a bias against urban agriculture dating to the colonial era. Here in the United States, the idea of what makes a city may echo the views of India and many African nations, but, from a physical perspective, our cities run the gamut: large or small, dense or sprawling, relatively new or relatively old, planned or haphazard, growing or shrinking.

The advantage of sprawling, low-density cities to urban farming are obvious: the market of a city with the land availability of a suburb, often in the warm and sunny South or West. Chattanooga's zoning ordinance permits a specific Urban Agricultural District, for example—even permitting animals, such as cows and ostriches—but the minimum lot size is 20 acres (8 hectares). That's quite a different scale than many older, denser cities could accommodate.

Some of the most interesting new developments are in old Rust Belt cities and other former industrial areas, such as New York and Chicago, Philadelphia and Pittsburgh, and Milwaukee and Cleveland. These cities' fortunes have taken different paths since the heyday of American industry, but it's in perhaps the hardest-hit cities—such as Detroit—that we see some of the most promising work.

Current Challenges and Opportunities

Unlike in China, where central planners had the ability to stretch out cities, or Cuba, where a real economic collapse caused an urgent need for solutions, most US cities have neither the urgency to roll out urban agriculture on a mass scale nor the authority to do so. Yet all is not lost. Just as those countries saw urban agriculture as a solution and worked within their national frameworks to promote it, so can we. Currently slowing the adoption of urban agriculture in the United States is a number of factors which are, if not roadblocks, at least speed bumps. Only in facing these challenges can we hope to make urban agriculture reach its potential.

Land Costs

Even if the federal government and municipalities were to completely ignore the value of urban agriculture and leave its development entirely to individuals, the growing crowd of enthusiasts could make it a reality if they could obtain the space to do so. And there's the rub. Land prices are likely the biggest limiting factor in urban agriculture the world over. Much can be accomplished on fire escapes, windowsills, and balconies, but optimal development of urban agriculture requires some acreage on the ground or in the air. The problem here is that those who would like to farm seldom own such property; if they rent it, they typically lack the land tenure necessary to make an investment in agriculture enduringly worthwhile.

Though it may seem counterintuitive, Greensgrow Farm in Philadelphia has found no problem with obtaining space to farm. In fact, according to its website (www.greensgrow.org), one of the farm's main lessons learned since 1997 is that "unlike rural areas, good land is readily available in cities, and it's cheap. This is because the criteria that define 'good land' are different in urban areas." As Greensgrow sees it, good land in an urban area is primarily brownfields. "Vacant, post-industrial properties in the inner city are, by definition, located right inside your market—both the immediate neighborhood and the larger city itself." These brownfields have no leasers because they are contaminated and no buyers because remediation and liability concerns make them too expensive to own. So this type of land essentially costs its owners money and generates no income. The key, according to Greensgrow, is to lease the land for a nominal fee and seal out the contaminants by, for example, using raised beds or containers. As a result, "one of the best uses for brownfields that cannot be immediately redeveloped is as an interim usage for urban agriculture."

This former galvanized-steel plant is currently Greensgrow Farm.

Another option is to exploit spaces that are undesirable or unusable commercially, such as road median strips and verges, rooftops that can bear light loads (think rooftop greenhouses) but not additional

stories, rights of way under power lines and along train tracks, and the yards and rooftops of churches and schools, many of which are already ahead of the curve. For instance, another urban farm in metropolitan Philadelphia, Somerton Tanks, is on a half acre of land owned by the water department right near (you guessed it) giant water tanks. Similarly, temporary spaces, such as vacant lots slated for development after a growing season or more, provide another opportunity for growers using mobile vehicles (such as barrels, EarthBoxes, Dumpsters, or bags of soil of any size) for growing.

Designating some areas as urban-agriculture districts may serve to protect some community gardens and farms, but it also poses the risk of the same kind of unintended consequences that can attend rent stabilization: by protecting some properties from the market, the total supply is diminished and the value of the remaining properties is driven up. In extreme cases, it could mean that the neighborhood farm remains, but not the neighbors, who are all eventually replaced by wealthier ones.

Imagine that an old downscale industrial zone is absolutely transformed through urban agriculture. Beautiful fruit trees line the streets, green veils of kiwi and grapevines obscure the walls, florist-grade flowers tumble along the walkways, and every available yard and rooftop puts forth a bounty of vegetables. You even suspect some basement mushrooms or aquaculture. All of this increases the disposable income of the farmers, attracting little cafes and restaurants. Unemployment rates go down, and the area's reputation goes up. Before you know it, the neighborhood's gritty charm attracts wealthier tenants, sending up rents and the value of real estate. Suddenly, property owners realize that the hard-won little gardens on vacant lots can very profitably be replaced by condos. And, of course, you can't eat a condo. A major challenge for many urban agricultural projects with social missions, especially those aimed at empowering underserved communities, is ensuring that the benefits of the farm primarily accrue to that community. Hilltop Urban Farm (see Afterword) is a good example of an initiative conceived not just for, but with, the intended beneficiaries.

While urban agriculture would not be the only factor in real life, this kind of transformation does happen. When Beijing began to permit peasant families to farm and sell their goods in a market-based fashion, new wealth prompted a boom in home construction—on agricultural land, which had already been in short supply. Beijing could fall back on its central planners to hammer out a compromise, but ours is a trickier situation. Then again, the flip side of our low food resilience is that we don't rely on urban agriculture, so that kind of occasional loss is unlikely to affect us.

Transactional Costs

Six hundred acres (243 hectares) of window boxes may be great for the United Kingdom, but it's hardly a recipe for food security. It can give a touch of food security to the box farmers, but it's tiny on a national scale. Furthermore, it involves millions of different growers in thousands of different communities with hundreds of different agricultural preferences. Even if all of that window-box acreage were somehow coordinated, there are still transactional costs: communicating the great window-box agenda, distributing seeds, setting up local markets, harvesting, getting the right amount of produce to market, and so on. The transactional costs of getting all this done most likely outweigh the sales value of an entire year's crop.

It's an exaggerated example, but it illustrates real problems facing small- scale farmers. They don't grow enough to make growing and marketing produce their full-time jobs. If several of them banded together into a cooperative, then perhaps they could all continue farming part-time, with one of them occupying the rest of his or her time marketing the produce for the whole group. Easier said than done. Farmers' markets are an option, but would they allow you to (or could you afford to) rent a stall just for a basketful of eggplant? So you end up with enough of a harvest to have a surplus but not enough of a surplus to make marketing it worthwhile. It's inefficient and one of the realities supporting the economies of scale pushed by Earl Butz with his philosophy of "get big or get out."

Perhaps the best response to the question of transactional costs is the point that in the United States, urban farming is not primarily about food, at least not in the sense of the naked delivery of calories. Rural agriculture almost certainly wins at that, particularly in the realm of meats and carbohydrate-laden grains. Yet that is the foundation of industrial (which is to say, mainly rural) agriculture. We need calories to live, and it delivers them at a profit. The motivations of urban farmers, however, often involve the education of youth, tangible productivity, reconnection with the earth, shared work, living beauty, and a sense of place. And you get some food from it, too. It has efficiencies of soul.

That being said, many farmers do grow enough to make selling at a farmers' market worthwhile for at least some of the year, but they may have difficulties breaking into these markets. When I visited The Bridge Urban Farm in the Bronx, for example, they were having a hard time finding a local retail farmers' market where they could sell their produce. At the same time, larger-scale farmers north of the Bronx, in the Hudson Valley outskirts of New York City, were not able to market their wholesale goods in the city's main market, Bronx-based Hunts Point—the largest produce market in the world—and were instead selling their produce through a market nearly 100 miles (161 km) away in Philadelphia. That's still kind of local, depending upon your definition of local food, but it's far from optimal. Help from municipalities and nonprofits in setting up farmers' markets, zoning laws that permit on-farm sales, and other such measures can help things move more smoothly and in fact are being slowly implemented in cities around the country.

Finally, the transactional-costs argument often falters, at least in cities with abandoned properties and brownfields. Such cities can support a whole gaggle of decentralized small-scale farms. Even stipulating that small, decentralized farms are

Even modest vacant plots can provide tremendous growing opportunities for city dwellers.

less efficient than large centralized ones—which many urban agriculturists would dispute—inefficient existing farms are still preferable to efficient nonexistent ones. If the land is unused and unwanted in the commercial market, then surely some productive use is better than none at all.

Policy and Zoning

Even if you have the land and a dedicated market, zoning restrictions can leave you dead in the water. This is particularly true for livestock. Zoning restrictions scuttled the livestock ambitions of BADSEED Farm in Kansas. Even Jennie Grant, who won a zoning victory for Seattle goatherds, is limited in how many goats and chickens she can keep. Beekeeping is forbidden in many cities. Zoning ordinances may be more restrictive for animals than for plants, but they still hold the power of life and death for urban farms larger than a balcony or cloistered yard. Urban agriculture needs a friend in city hall.

As Martin Bailkey and Jerry Kaufman note in *Farming Inside Cities: Entrepreneurial Urban Agriculture in the United States*: "Local government obstacles center around issues of policy and practicality (urban agriculture being a nontraditional land use), and attitude and ideology (whether urban agriculture represents the 'highest and best' use of city land)." Bailkey and Kaufman may have a more generous view of urban agriculture than Joel Kotkin, but on one thing they agree: many cities are losing middle-class residents to the suburbs.

When cities consider their development plans, they need to face reality: those pockets of spare land that they want to turn into research centers or shopping malls or upscale apartments may never see such a use. Bailkey and Kaufman see it this way, too: "It is true that some city vacant lands, because of their strategic location near the center of town or near a river or lake, will be attractive sites for future housing and business developments. Yet considerable land in these cities will likely continue to be vacant, unsightly, and unproductive."

Exhibit A: Detroit. This is not the fate of every city, of course. The Malibus and Palm Beaches of the country probably don't need to worry about high unemployment rates and trash-strewn lots. For many of the cities between these extremes (which is to say, almost all of them), earmarking some development for urban agriculture makes sense to beautify the city, mop up underutilized land, increase property values, and provide jobs, all while building resilience, if not complete redundancy, into the local food system.

In the time since the publication of *Farming Inside Cities* in 2000, many cities have begun to allow and even encourage urban agriculture through a variety of

tools, such as distribution of idle municipal land, favorable changes to zoning restrictions, development of farmers' markets, and establishment of food-policy councils. Because urban-agricultural enterprises often advance the "triple bottom line" of people, pockets, and place, they are a natural fit to advance cities' broader goals for sustainability that have gained currency in recent years. In 2017, urban farmers everywhere gained a huge ally with the launch of the searchable Global Database for City and Regional Food Policies. The new database was launched by the Resource Centres on Urban Agriculture and Food Security (RUAF) and the University of Buffalo's Food Systems Planning and Health Communities Lab (the "Food Lab"). The Food Lab also maintains a database of policies within just the United States and Canada, which is a great first stop to find progressive policies in cities comparable to your own. (See Notes & Resources for both databases).

In addition to zoning practices favorable to urban agriculture, cities can encourage its development in many ways, such as tax incentives, purchasing power (for example, requiring that public schools procure 20 percent of their food from within 20 miles [32 km] of the city), technical assistance, and possibly help negotiating for collective insurance coverage or bulk discounts for agricultural supplies. Some municipalities provide discounts on water for urban agriculture, and there are many ways that cities could help with recycling compostable waste, simultaneously reducing a city's waste stream and providing urban agriculture with compost that can be used to grow food directly and/or be sold to others.

How Much Is Enough?

If urban agriculture were to be embraced on a federal and local level and to really blossom across the United States, how much of our diets could it comprise? By some measures, it already produces more than half. "While farming is relatively more important in nonmetropolitan areas than in metropolitan areas," notes one report of the Trade and Agriculture Directorate of the Paris-based Organisation for Economic Co-operation and Development (OECD), "a majority of farm production occurs in metropolitan areas of the US."

In 2005, more than 56% of the gross domestic product (GDP) from farming, forestry, and fishing was produced in metropolitan areas. This directly contradicts our experience in most cities and demands a lot of qualification. Much of the counterintuitive numbers have to do with vagaries of how the Census Bureau and others distinguish metropolitan and rural areas—it's no easier for them, apparently, than it is for the rest of us. The report notes, for example, "Based on the Census 2000 population and the current Core Based Statistical Area classifications, 51% of *rural* residents live in *metropolitan* counties."

Huh?

And it's worth noting that the figures measure the dollar value of production, not the calorie value. High-dollar-value short-lived items, such as salad greens and fresh tomatoes, are likely to typify metropolitan agriculture, while low-dollar-value long-lasting staples, such as corn and wheat, are more likely to be rurally produced. Considered from a calorie breakdown, I'd bet that at least two-thirds of agricultural production occurs in rural counties, however strangely the Census Bureau defines them. Let's not forget, either, that metropolitan areas cluster on the coasts, so they have a near monopoly on the value of fishing production. This figure also aggregates metropolitan agricultural production into one lump sum.

Sure, X percentage of the food you eat may be of metropolitan origin, but what are the chances that the origin is your metropolis? It might technically be urban agriculture, but it's not necessarily reflective of the self-sufficiency envisioned by that term. Moreover, about half of our agricultural production by value is exported. So it might be technically urban agriculture here, but it's consumed abroad, which again runs counter to the self-sufficiency goal.

It's hard to predict an increase in urban-agricultural production without having an accurate baseline. I would guess that using an arbitrary but fairly commonsense definition of a city—the official municipal boundaries plus immediately adjacent counties—the average city in the United States produces maybe 5 percent of the food it eats. If that's fair, I'd consider an average of 15 percent "enough"—about one day's worth of food every week. That provides a significant backstop against dramatic and unexpected food shortages and would probably employ about as many people as would want to be urban farmers. I'd consider 20 to 30 percent ambitious but possible. Some cities would never quite get that high, and some (maybe Detroit) could even top it, but there's a limit as to how much local food most people would want to eat. Cereals, beef, cow's milk, pork, and lamb are likely to remain outside of the city's domain, not to mention exotics such as coffee, chocolate, bananas, and other fruits out of season in our part of the world. If you include another ring in this bull's-eye model (in other words, the counties encircling the counties immediately adjacent to the city), a city's level of self-production would probably kick up 5 percent or more.

The Future of American Urban Agriculture

If there's anything that can be said accurately about prediction, it's that it is often wildly inaccurate, more so the further forward it looks. However, it appears very likely that urban agriculture will intensify. The current momentum in urban agriculture suggests several trends, including development of proven models, expansion of for-profit and corporate interest in urban agriculture, and an identity crisis.

Development of Proven Models

Since the beginning of the twenty-first century, urban agriculture in the United States has mostly been an enthusiastic jumble of amateurs in the best sense of the word: people doing work that they love—some with better agricultural expertise, some with keener business sense, some with greater resources at hand, but none following a template for the professional pursuit of American urban agriculture. It simply hasn't existed, except with urban community gardens. Unlike community gardens, notes Bailkey, which have an accepted model, "there's no paradigm for most of what we're talking about when we talk about urban agriculture." Among today's amateurs are the pioneers who will be emulated by tomorrow's professionals. They will be what Mary Seton Corboy, the CFH ("Chief Farm Hand") of Greensgrow Farm, calls "standard bearers of the profession." We don't see them yet, but we will.

It's likely that several major models for urban agriculture will emerge, not all suited to every city. If we look to cities around the world with very robust urban agricultural communities, however, they seem to share five elements in common: official encouragement, central coordination, a multipronged approach, integration of new technology, and some aspect of recycling (such as food waste or gray water).

Official Encouragement

While it's hard to measure official encouragement, I think it's safe to say that advanced states of urban agriculture are never reached in cities where urban farmers work at odds with city hall. Almost always, if not always, urban agriculture progresses in places where the municipal government actively works to foster the practice, such as Shanghai (surprise!), Singapore, Vancouver, or even Madison, Wisconsin. The good news for guerilla farmers in recalcitrant cities is that official attitudes can change. They did in Kampala, Uganda; they did in Havana; and they are changing across the United States right now, as evidenced by the laundry list of cities with new urban agriculture ordinances listed in the introduction to this book. Nothing quickens the turnaround in attitude like the success of similar cities—except, perhaps, absolute necessity—so we can expect more cities to welcome it one way or the other.

Central Coordination

Official approval may be a prerequisite of wide-scale adoption, but central coordination is a major advantage. A thousand acres of rooftop radishes is impressive but inefficient. What do you do with them all? Central coordination greatly facilitates things by making the most of different crops in different spaces. Municipal government often plays this role internationally, and it's no secret that some of the best coordinated cities are run by governments fairly described as authoritarian. This is one of those instances where our political advantages are disadvantageous practically. Fortunately, central coordination need not be government-based; in the United States, nonprofits often play a coordinating role in conjunction with varying degrees of official encouragement. Since this requires cooperation, the success of such public/private partnerships will probably vary greatly by city.

Multipronged Approaches

Many cities in which urban agriculture provides significant food security take multipronged approaches. Elements often include provision of farming space around new housing developments, school-based gardening programs, designated urban-agriculture districts divided by type (aquaculture, hydroponic vegetable cultivation, and so on), technical assistance, seed-distribution programs, and agroforestry plans. As with official encouragement and central coordination, a comprehensive multipronged approach to urban agriculture is relatively easy in a benevolent dictatorship but not so much in a democratic, market-based system. You can't just retrofit an existing city to be Eden. Although few, if any, American cities have the right combination of stakeholders and the right degree of power to implement a true multipronged approach, they are on the path to do so. Plans such as Portland, Oregon's "Diggable City" (2005) report and "FoodNYC: A Blueprint for a Sustainable Food System" (2010) outline visions for cities' futures, while food-policy councils such as those in Toronto, Cleveland/Cuyahoga County, and Seattle/King County bring together stakeholders as well. As cities draft their periodic master plans, urban agriculture is likely to enter the picture on a rolling basis. The cities taking the most comprehensive action today, of course, will likely also be those best positioned to enjoy the many benefits a decade from now.

Technology

Although "urban agriculture" still conjures up an image of the most basic, earthy work, leading cities often embrace technology. Vegetables are grown outside in hoop houses and inside in climate-controlled greenhouses, often hydroponically (pictured). Aquacultured fish swim in recirculating tanks. Xiaotangshan Modern Agriculture Demonstration Park in Beijing, for example, uses GPS and geographic information system (GIS) technology to monitor processes such as watering and fertilization. Jac Smit was excited about technology that measures the tension in trellis wires to best estimate yields; if it's not already used somewhere in urban agriculture, it probably will be soon. Aquaponics is often seen as the future of feeding cities, and doing so efficiently—and without pollution—demands a high level of technological know-how. Corboy goes so far as to say:

> **No one has really invented any of the tools for urban agriculture. We still use a modification of everything that's already out there, somewhere between the home consumer and the small farmer. Once somebody comes up with some kind of tools that make sense for urban agriculture, then it's going to change the nature of what urban agriculture is, also—just as it changed the nature of what agriculture was at the end of the eighteenth century.**

We may not have achieved the kind of vision Corboy envisioned yet, but we are getting close. It is easy to find online (if not in stores) LED lights that are fine-tuned for indoor growing, a wide variety of automatic watering systems, and a huge array of growing vehicles, from small herb-sized pots to ready-made raised beds to entire modular farms built into shipping containers.

Recycling

Finally, many cities with large-scale urban agriculture programs utilize some kind of recycling. Each year, about 130,000 tons of food waste find their way into pigs in Hong Kong, which incidentally produces about 15 percent of the city's own pork. In Milwaukee, cultivating compost is one of Growing Power's chief tasks, and it cultivates a lot: more than 2,000 tons of food waste, 500 tons of brewery waste, and 20 tons of newspapers and coffee grounds each year. The decomposing compost helps heat greenhouses; the finished compost provides a fertile medium for its raised beds; and the extra worms from its vermicomposting operation become food for its tilapia and yellow perch. The fishes' wastewater then fertilizes growing beds. In Salt Lake City, a patch of unused city land is being used to grow safflower, fertilized by biosolids produced by wastewater treatment. The safflower is intended, in turn, to become biofuel for the municipal fleet.

Although not every leading city incorporates all of the five elements that seem to be common to urban farming success, these elements are mutually supportive and go together naturally. Enthusiastic municipal governments can direct the sewer authority to, for example, divert some untreated water to a timber-producing forest within city limits, sell treated water at a discount to city farmers, and distribute biosolids freely to gardeners and farmers. This

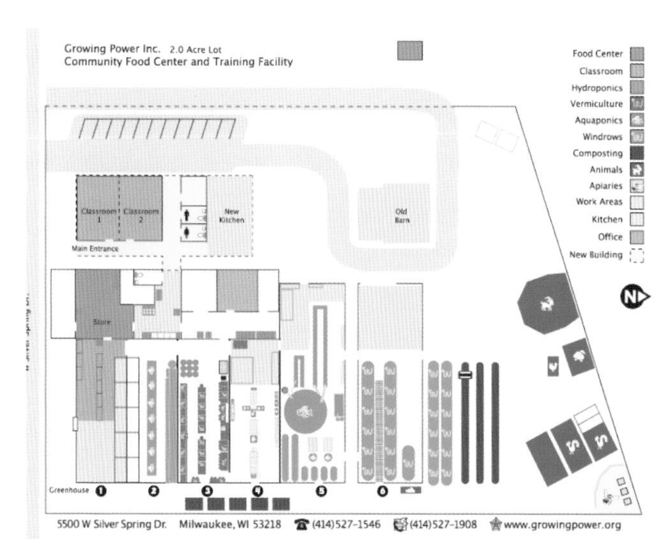

Growing Power's headquarters is a model of efficient use of space.

alone would reflect all five elements. If only it were that easy in practice. The ASU/Google study mentioned earlier estimates that existing vegetation in urban areas worldwide (which includes both urban agriculture and other vegetation) already diverts about 50 billion cubic meters of stormwater runoff annually and sequesters between 100,000 and 170,000 metric tons of nitrogen through the use of wastewater, nitrogen fixation by plants (such as peas), and other means.

Growth of Commercial Enterprises

As urban farming models develop and demonstrate success, profit will increasingly join the many other motives for urban farming. Even now, it is far from an entirely nonprofit activity. Many urban farmers hope to make their livelihoods that way, and a lucky few do. SPIN-Farming and other methods exist for this purpose. Dozens, if not hundreds, of businesses make money selling products, such as containers and tools, that find their way into urban food production, and restaurants and markets are already underwriting local urban efforts for the cachet that urban farming brings. All of these will no doubt multiply.

The real change, however, will involve corporate growing efforts. A handful of for-profit companies have attempted and later abandoned urban-agriculture initiatives, but the same forces that are coalescing to make urban agriculture attractive to individuals can make it more attractive to companies, too. They have a growing market.

Tomator producer Eurofresh Farms has the largest greenhouse facility in the United States (over 300 acres [121 hectares]) and is located within 85 miles (137 km) of Tucson. The world's "largest sustainable indoor fisheries" and "largest indoor producer of tilapia," Virginia's Blue Ridge Aquaculture, lies within 60 miles (97 km) of Roanoke, Virginia, and Greensboro and Winston-Salem, North Carolina. Both companies ship their goods widely now, but it's easy to imagine competitors settling down closer to larger cities. Or even *in* some of them.

The boldest venture of the moment is Hantz Farms and its ambition to become an agricultural Disneyland. Attracting visitors is not an outrageous concept; Xiaotangshan Modern Agriculture Demonstration Park was designed with agritourism in mind and attracts tens of thousands of visitors per year. In fact, the park's attendance income now exceeds its production income, which is considerable.

Identity Crisis

Another near certainty in the future of urban farming is a crisis of identity. Many longtime observers hesitate to say what urban agriculture is, particularly amid its surge in popularity. Corboy finds all of the professed motives for urban agriculture almost humorous, and Michael Levenston of City Farmer believes that "it'll shake down when it's no longer fashionable." Business interests may well end up driving a shakedown. Precisely because urban agriculture means so many different things to different people, the entry of large corporate players threatens to fracture the affinities among the advocates of local food, self-sufficiency, food security, skill building, youth development, antiagribusiness, anticorporate, and other agendas who now champion mostly small-scale urban agriculture as a way to advance their individual missions.

Corporate players would make it much easier for locavores to eat local food, for example, but this could come at the cost of the "know the farmer" ethic. Opponents of industry agriculture will have to decide whether they object more to the agricultural practices of agribusiness or to its corporate nature. Will assembly-line hydroponic agriculture satisfy organic traditionalists? How would corporate competition affect all of the urban farmers who entered the field precisely to move from being consumers to being producers?

Whatever their more specific motivation (such as jobs, youth development, or community building), these urban agricultural groups are generally different than corporate farmers in a fundamental way. Says Bailkey:

> **There is a different mindset behind looking at urban agriculture the way it's largely been looked at...as a grassroots, from-the-ground-up, community-based activity versus an opportunity to invest venture capital. It's two completely different mindsets. Probably the biggest difference is that I don't think the venture capital people are as interested in the process, except to make it more efficient. For the grassroots side, it's both the product and the process...**

In other words, it's not just the food—it's whom you engage, what kind of community you build, and so on. Those are not fundamental corporate interests.

It may very well be the case that no big corporate farms find their way into cities—Corboy, for example, sees development within 75 miles (121 km) of cities as more likely—but there's a good chance that they will. In many leading urban-agriculture cities, consolidation of operations and industrial intensification are key elements of their overall strategies. Corporations may not be the entities that drive these changes in those foreign cities—often it is government authorities—but their most reasonable counterparts in the United States are corporations. Few community-development organizations are in the position to buy hundreds of acres of municipal land, negotiate with city hall about water access, or build a $200-million vertical farm, but many corporations are. And, because so many nonprofits have been engaging their constituencies in urban agriculture, there's a good possibility that they could find a qualified workforce without too much difficulty.

What effect would a large for-profit farm have in a city with a thriving group of community gardens, small entrepreneurs, and nonprofits devoted to urban agriculture? There's some fear, in Detroit at least, that a corporate farm would be disastrous. Small farmers simply couldn't compete. As local produce would become much more available, urban-agriculture-focused nonprofits— which are able to raise money, in part, because Detroit is so well known as a food desert—would find their funding drying up.

Yet this scenario is not inevitable. There's a good chance that longtime urban farmers could find skilled work with corporate growers, for example, and Mike Score of Hantz Farms points out that they would most likely market at the wholesale or large retail level, leaving the majority of the consumer retail market for other urban-agriculture activities. Professor Hamm acknowledges the tensions but also senses "more recognition now that there's room for all of these different players to coexist." In the proving ground of Detroit, this coexistence may very well produce the most successful models for urban agriculture, small and large, nonprofit and for-profit, and otherwise.

Ways to Help

Even if your interest in urban agriculture is academic and you have no intention of scratching dirt yourself, there are many ways you can help improve urban sustainability in general and urban farming in particular.

Help Local Urban Farmers Directly

If you have extra space but no interest in agriculture, letting a local urban farmer use your space not only benefits him or her but also probably results in delicious dividends for you. If you have a decent-sized shed, you might even offer to host a local tool library from which local farmers could borrow hoes, shovels, rototillers, and other items they need.

Get Involved, but Keep Your Hands Clean

Don't dismiss the power of an informed and passionate citizen. Write to your local council members and the mayor, encouraging the development of urban agriculture. Write op-eds for local papers. Attend town meetings. When talk of redevelopment and other urban agriculture-related matters is in the air, volunteer for a board or

Even if you don't farm yourself, you can enjoy urban produce through a CSA or farmers' market.

committee. Ask what your local cooperative extension is doing to support urban agriculture, and if the answer is "not much," find out why. Write to your local community foundation and ask it to support urban agriculture if it is not already doing so. Follow the developments in similar but more agriculturally progressive cities, and keep a file of potential cost savings and other benefits achieved (or anticipated) through urban agriculture. Pressure your city to keep up with the Jonesvilles of the world. Prepare a thirty-second "elevator speech" that extols the benefits of urban agriculture to use on strangers. Every time you speak up, you're furthering the cause.

Recycle

Deploy your persuasive powers to encourage your city to compost food and other appropriate waste. This may or may not benefit local urban farmers, but it at least reduces disposal costs (and attendant greenhouse gases) and preserves the nutrients that would otherwise be lost. If citywide composting is not a possibility, aim for city-subsidized distribution of recycling paraphernalia. Seattle, for example, offers rain barrels and yard-waste composting bins at subsidized prices, as well as Green Cones, solar digesters manufactured by Solarcone, which efficiently compost food waste. Furthermore, your city could also offer subsidized EarthBoxes or other growing containers. And if you're not already, start recycling and composting yourself.

Support Local Agriculture

Whether urban or rural, most farmers face a tough and often thankless job. You can show your appreciation for their efforts and local green spaces by buying their goods. Farmers' markets provide a great way to do so, as do CSAs. Local Harvest (www.localharvest.org) is a great starting point to find these outlets in your area. If you're still in activist mode, try leveraging the buying power of large institutions. If the local public school system (or hospital, or department of corrections, or S&P 500 company) decided that 20 percent of the food it serves would be local, however defined, it would provide enormous opportunities for urban farmers to succeed while delivering all of the benefits attributable to local agriculture.

A Grain of Salt

If you plan on jumping into urban agriculture, start small and don't quit your day job just yet. You might join a local CSA or try some container gardening first. Don't bet your proverbial farm on a real one in the city. Martin Bailkey talks about "optimism within limits" in regard to urban agriculture writ large, and the same attitude probably behooves the would-be farmer. Just ask Mary Seton Corboy, who laughs a lot about her experience as an urban farming entrepreneur but notes that there were many tears early on (but a lot of laughter, too).

There will always be the need for food in cities as well as a means of producing it. In a world where so many familiar tangibles such as books, classrooms, DVDs, and CDs have migrated into the digital ether, it's good to know you can count on the solidity of a potato and the bite of a radish. "There's no iFood," says Levenston of City Farmer. Food will always remain something that you can see, smell, taste, touch, and—if you desire to—grow. The first step is to learn a little bit more about how.

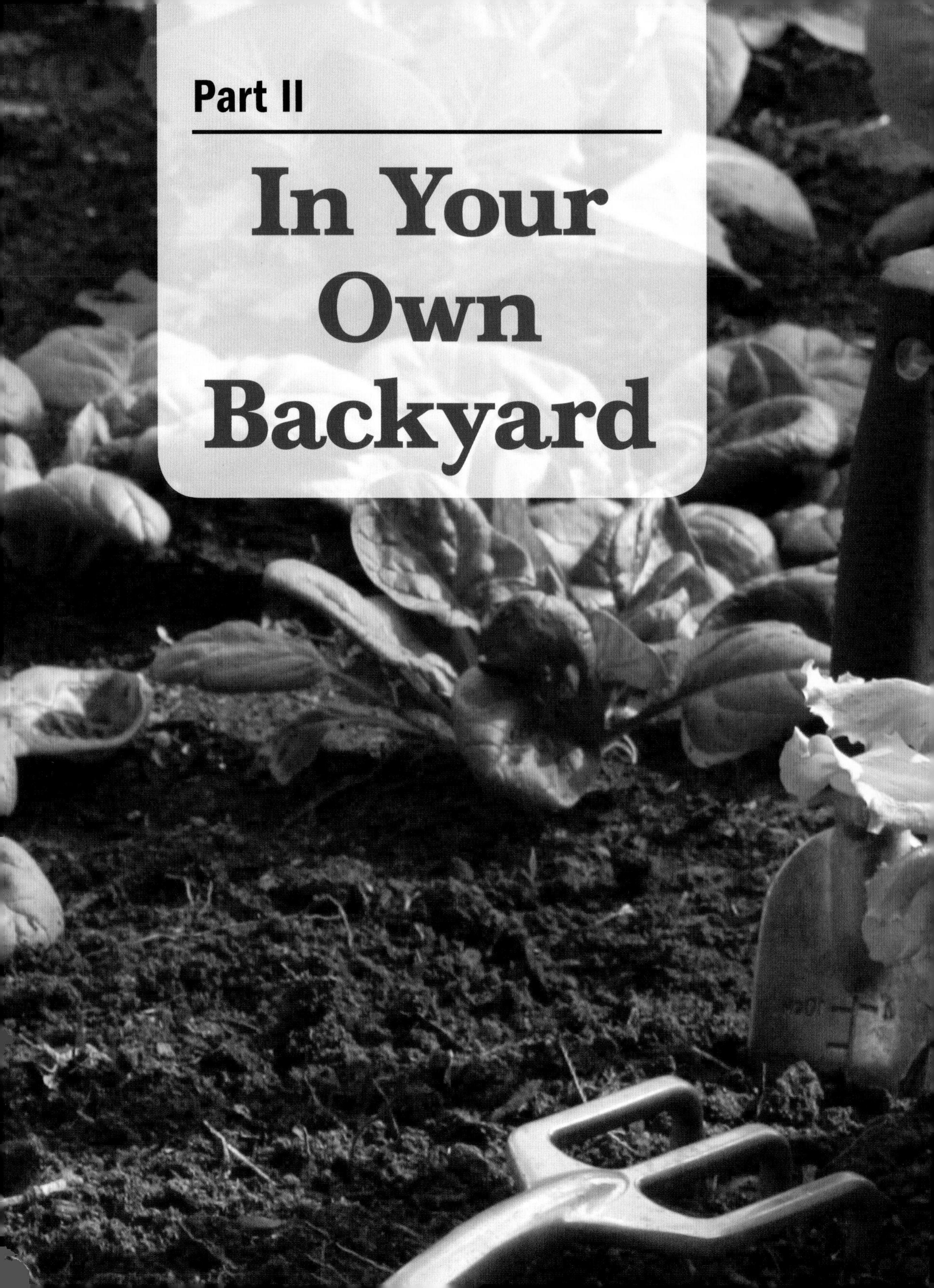
Part II
In Your Own Backyard

CHAPTER 4

Starting Your Farm

OK. You're intrigued. You wonder if homegrown tomatoes really do taste better. And you want to stick it to that supermarket produce manager who always gives you the evil eye for smelling the peaches. So how do you become an urban farmer?

The first thing to do is decide what kind of urban farmer you will be (a "good one" is a given). You can find whole documents devoted to "typologies" of urban farmer, but let me brutely reduce them to three main species: subsistence, recreational, and entrepreneurial.

A subsistence farmer grows to make a living, usually choosing crops or livestock to meet his or her own food needs or high-value items that can be sold profitably to others, whether food, flowers, or other products. As Dr. Martin Price of Educational Concerns for Hunger Organization (ECHO) has noted, hunger among the world's urban poor usually reflects a shortage of income rather than a shortage of food. Since food is often their single greatest expenditure, subsistence farming can reduce the biggest category of expense and possibly even generate additional income. It's a win-win situation. Most of the world's estimated 800 million urban farmers fall into this category. Examples might include a mother growing corn and sweet potatoes at a scrap yard in Kampala, Uganda, or a Philippine couple on the outskirts of Manila growing jasmine flowers to weave into garlands.

Detail of the vegetable garden, or *potager*, on the property of a French country house.

Recreational farmers grow out of interest rather than necessity, although they may subsist on what's grown by choice, not because they have to. The primary audience is the farmer himself, so the choices of crop are abundant. I also include here people farming out of sustainability concerns or local-food interest; their motivations may not be recreational in terms of entertainment value, but they are not at risk of starving if they do not grow food to eat or sell. Traditional French kitchen gardens (potagers) would most likely fall into this category, as would most urban farms in the United States. So would typical community gardens, which are akin to group recreational farms. The Urban Ministry Center Community Garden in Charlotte, North Carolina, for example, provides fresh vegetables and fruit for its soup kitchen. It may produce only a fraction of the ingredients for the 200-plus meals that the center provides each day, but the garden presents the opportunity for its homeless neighbors to garden, interact, and just be—a place where, according to its blog, "... our harvests of stories, good will, and shared experiences are every bit as important and nourishing as our tomatoes, okra, broccoli, and greens."

The entrepreneurial farmer is motivated primarily by making profits. Driven by the market, he or she is likely to choose high-value, perishable items such as gourmet greens or other specialty produce, fresh flowers, or live aquacultured fish. Although the capital inputs required to produce those items may be higher in the city, the reduction in transportation, refrigeration, and storage costs help make them more competitive with imported goods. In this way, the entrepreneurial farmers and subsistence farmers who market their goods are very similar in terms of motivation (to make a living) but very different in scale, with entrepreneurial farmers likely to have larger farms and formal business plans.

Entrepreneurs are almost certainly the smallest category of farmer internationally, as well as in the United States, but may hold the largest potential impact because of their capacity for growth and innovation. A good example of this is Sky Vegetables, the previously mentioned start-up that developed the technology to build hydroponic farms with year-round production on urban rooftops.

Subsistence. Recreational. Entrepreneurial. These are not inviolable categories. The dependence of subsistence farmers on their homegrown food may vary with the season or economic climate, while social or community concerns may figure prominently in the motivations of entrepreneurial farmers. Recreational farmers might earn a decent income from their pet projects. And so on. A 2015 study by Carolyn Dimitri et al found that many urban farmers have a tangle of motives, not just one or two clear-cut ones. Nevertheless, having a general idea of which category you fall into is helpful for planning.

Take crop choice. A recreational farmer might plant something notoriously unreliable, or simply unusual, just for fun, whereas a subsistence farmer lives too close to the edge to take that chance and would do better by picking fail-safe crops. Entrepreneurial farmers can weigh the risk of failure against the possibility of higher profits for something hard to find. This summer, for example, I planted okahijiki (Salsola komarovii), a land-based plant eaten in Japan that is similar to hijiki seaweed. It actually grew pretty well, but I had never tried it before—or the original hijiki, as far as I knew—so I had no idea how it tasted or who, if anyone, would like it. It would have been a phenomenally risky choice if I were a subsistence farmer, and I would have had to do more research on local demand if I were an entrepreneur (unless, perhaps, I lived in Japan). But I'm just a recreational farmer, and I have the luxury of experimenting.

Okahijiki: like seaweed for landlubbers.

Urban Farmsteads

Once you have an idea of the urban farmer you hope to become, the next thing to do is stake out your farmstead: a balcony or small space, a rooftop, a lot, or a network of sites. If you don't own the desired site, you will need to obtain the site owner's permission, which means you'll need to know who the owner is. If you don't know (and can't find out from the building manager or lease agreement), contact the city or county clerk; you'll probably want someone called a tax assessor, registrar of deeds, or something similar. The ownership information is a public record, so don't be afraid to poke around. The information might even be accessible online at the municipal or county website.

The University of California Cooperative Extension (UCCE), Los Angeles, produced an excellent Community Garden Start-Up Guide that provides advice on the process of organizing neighbors, investigating a site, obtaining a lease and liability insurance, and planning your garden. The guide even includes a sample contract for participants to sign. You can find it and other community garden start-up resources at the American Community Gardening Association's website (https://communitygarden.org) under the "Resources" tab.

Breadbasket to Balcony

If you're measuring your back forty in inches—from windowsill-size to a small balcony—containers will be your bread and butter. Fortunately, containers are marvelously diverse, and even these small farmsteads give you terrific opportunities for container gardening, including vertically intensive techniques that stretch out your domain. A sunny indoor window provides a good climate for herbs such as basil, cilantro, and thyme. Watercress might grow next to them in a jar of water. Chinese yam (*Dioscorea batatas*) could vine up one side of the window, with an upside-down cherry-tomato planter hanging on the other side.

Combining lettuces of different colors can produce an edible bouquet, as shown by this example at an urban farm in the Bronx run by The Bridge.

An outdoor planting box at the same location could accommodate arugula or spinach, beets or onions, purslane or rosemary. Mixed-color lettuces can be gorgeous. If it's legal to use in your city, a fire escape could hold a few potted tomatoes or peppers on the landing and edible nasturtiums or chrysanthemums hanging in baskets from the bottoms of the stairs going up to the next floor.

A balcony might host potted fig and dwarf lime trees, a cucumber vine trailing along the wall, ornamental kale in containers, and buckets of peas and eggplant. Edible sprouts (such as alfalfa), microgreens (such as arugula), and shoots (such as pea tendrils) can be grown year-round, even under artificial light. For such small spaces, the choices are far more numerous than you might at first think. You are unlikely to subsist on your crops at this scale, but you can at least impress guests, cut down on your food bills, and reduce your ecological footprint.

Rooftops: The Frontier above Us

There's no image of urban farming quite like green roofs, which ECHO's Dr. Price has called "the frontier above us." In built-up cities, rooftops provide a tremendous surface area, often with good access to the sun.

Despite their many advantages, green roofs pose several challenges, the first of which is access. That's easy if you own the building, but something to be negotiated if you're a poor schlub like the rest of us. Owners may have legitimate concerns about rooftop gardens, including safety, liability, and the possibility of foot traffic damaging asphalted surfaces and leading to leaks. Be prepared to launch your counterarguments as you need them: an agreement to limit access to named persons, willingness to sign a waiver of liability for any accidents, an offer to line a footpath with protective material to avoid damaging the surface, the promise of heirloom tomatoes, and so on.

The owner is probably also the person most likely to know offhand the load-bearing capacity of the roof and where supporting walls are. Soil and water are heavy, so you want to be sure that the roof can support them if you have very ambitious plans. If you have more modest efforts in mind, a good strategy is to place containers along the outer walls. These stretches of roof are more likely to be able to bear loads than a random location in the middle, and the parapets can provide protection from strong winds.

Even if you have access to the roof, another key factor is water. You'll need it. A tap on the roof is ideal, but we seldom face ideals in life. Chances are, you'll have to lug it up yourself. If you're not made for that kind of exertion, you may need to consider another location for your garden. If you're up to the challenge of hauling water at more than 8 pounds per gallon, you might want to consider plants and planting strategies that conserve water and maximize the use of rain.

If your rooftop plans include greenhouses, then you really need to own the building or work something out with the owner. That's not a fly-by-night investment, and the greenhouse structure would probably need to be physically anchored to the roof to eliminate the chance of high winds blowing it into midday traffic.

Lots

In some countries, urban farmers can legally "squat" for a season on land that is not being used productively. That's not how it works in the United States. As with rooftops, farming lots is a cinch if you own one, but you probably don't, which means that you need to find the owner or manager and present your case. If the lot is bigger than a king-size bed, your case would probably be strengthened if you

The Bog

The challenges of farming could drive a man to drink. Lucky for Bob Philbin, an urban farmer in Scranton, Pennsylvania, he owns one of the city's most popular bars, The Bog.

Bob's farm is practically a one-person demonstration piece for urban farmsteads and methods. He starts seeds in a little greenhouse in his backyard, growing them on the roof of his bar and on a lot he borrows in exchange for fresh vegetables and paying the water bill. For convenience, he grows everything in EarthBoxes—about twenty on the roof and a hundred in the lot. Among their many selling points (including portability and water conservation), it can't hurt that his mother, Molly, was the company's education director.

Bob might seem to have it easy, yet all has not gone smoothly for him. He grows a nice variety of crops—including broccoli, cauliflower, sugar snap peas, eggplant, zucchini, Brussels sprouts, beets, and all kinds of beans—but he has found that the cafes and restaurants he supplies typically want just two things: lettuce and tomatoes. To sell more of his harvest, Bob started "produce Tuesdays" at The Bog, during which his regulars can find a wide range of seasonal vegetables on tap. Even with this additional revenue stream, the initial outlay for the EarthBoxes, organic growing medium, and fertilizers has made a real profit elusive so far.

Oh, and then there are the bees. Bob started keeping two hives at a friend's place in the East Mountain part of town, only to have them vandalized...by a bear. Bears may not pose a threat in your city, but Bob's next beekeeping challenge might: zoning laws that limit how close a beehive can be kept to other buildings. For now, they are settled on the apparently bear-proof roof of The Bog.

Bob has faced these challenges with good humor and keeps on looking out for new sites to farm. Old city lands and railroad sites seem promising, he says, especially since his container-based methods don't require fertile ground.

I asked Bob if he had any advice for would-be urban farmers. "I don't know. I've just kind of 'wung' it," he says, laughing. "Stick with it," he adds, "the markets are out there, and people are interested in local organic food."

formed the nucleus of a community gardening group and strategized a pitch before you approached the landholder. If your building's lot has unimproved land upon it, there's a chance that the landlord would view permission to farm it as a no-cost way to build goodwill among tenants and to make the site more attractive in the process. You should certainly present it that way.

If the lot is unconnected with the building in which you live, you may need to make more creative arguments. Local goodwill is an obvious benefit, for example. Beautifying the site can cut both ways: if the owner is looking to sell the lot itself, a community garden might dissuade potential buyers who worry about having to "evict" local farmers—not a selling point. If the owner is trying to sell an adjacent building, however, or lease units in that building, the next-door garden might be a positive development. Assess the situation and use your best judgment. If the lot is vacant and in poor shape, you might offer to clean it up as part of setting up a garden and suggest that your community gardening group would effectively patrol it. If dealing drugs or burglary is common in your area, the idea of committed locals looking out for the property could be a strong motivation to let you use it. And, of course, giving the site owner his or her pick of fresh produce is a strong selling point.

If you sense that the owner is on the fence but worried about the ground being disturbed or having to eventually cut down fruit trees and the like, you can use container gardening to your advantage. Garden State Urban Farms, for example, knew it could only use its original site in Newark, New Jersey, temporarily, so they planted everything in hundreds of EarthBox containers. In fact, no matter what farmstead you use, containers are often a great option.

Networks

One of the most clever solutions for urban farmers is to farm on several small sites rather than all in one location. For example, Wally Satzewich, a cofounder of SPIN-Farming, farms half a dozen sites in Saskatoon, Saskatchewan; another one north

of the city; and half a dozen more in the outlying village of Pleasantdale. Altogether, Wally's Urban Market Garden totals less than an acre. A network of beehives is a great example of agriculture that lends itself to geographic dispersion—and two organizations, the New York City Beekeepers Association and the Gotham City Honey Co-Op, thrived in the Big Apple even before beekeeping became legal there in March 2010.

There are many advantages to networked farming operations. Besides using space resources that would otherwise go untapped, they capitalize on the many microclimates that exist in cities. San Francisco's Sunset District might be foggy, windblown, and cool, for example, at the exact moment the Mission District is sunny, hot, and still. Different crops prefer different microclimates. Cities themselves tend to be warmer than surrounding land and offer many windbreaks, which can make production more reliable for some crops. Satzewich told me about one year when an early frost (in August—this is Canada, after all) wiped out rural green bean crops. Protected by just a few extra degrees of warmth, his urban-grown crop survived.

Whether you grow on a fire escape, on the roof, in a vacant lot, or at a network of sites, you may not feel like you have enough under cultivation to call yourself a "farmer," but bear in mind that each acre (0.4 hectare) of urban agriculture may produce up to thirteen times as much as a rural acre. So a measly eighth of an acre in the city is like a respectable acre and a half in the country, and an urban windowsill is like, well, thirteen rural ones.

Urban Farming Vessels

Once you have an idea of your urban farmstead, you can better choose the types of growing vessels you want to use: containers, raised beds, rooftop wick gardens, or the ground itself.

Containers: The Urban Farmer's Workhorse

The simplest patch of farm you can have is a humble pot. However, there has been an explosion in quite fancy but reasonably priced containers for the urban farmer. EarthBox, EMSCO Group, and other companies offer self-watering containers ideally suited to urban agriculture. They only take up about 2.5 square feet (0.25 sq m), allowing them to be used anywhere from the corner of a fire escape or terrace to grouped en masse on a rooftop or empty lot. And those are just basic workhorse containers: you can now also buy stackable plastic planters (see photo on page 95) that can accommodate anywhere from six to a dozen or more plants in a tower, or many-pocketed planters that can turn a wall into a vertical farm, or bottomless grow pots that you can push into a bag of soil. The options are myriad and increasing all the time.

If you'd like to consider your options before buying commercial products, however, read on.

Container Considerations

There are three main considerations in choosing containers. The first is drainage. Unless you're planning on growing aquatic plants, your container will need adequate drainage so that air can get to the roots. Assuming you already have good, porous soil, the standard approach is a hole in the bottom of the container. Most store-bought pots will include them already, though there are also uses for drainless containers—if, for example, you have a window box over the entrance to a building or want to rest a plant on good furniture. In those cases, a holed pot is usually inserted into the drainless one so that the plant's inner container can drain freely, with the outside pot catching the overflow. Coarse material such as pea gravel or broken terra-cotta at the bottom of containers, both holed and drainless, may help with aeration and management of extra water.

The second consideration is container volume. Here, you'd actually do well to follow former Secretary of Agriculture Earl Butz's advice to "get big." In general, the greater the volume you have to work with, the better. A conventional rule of thumb is that vegetables require a soil depth of at least 8 inches. A good working size for larger plants is 5 gallons

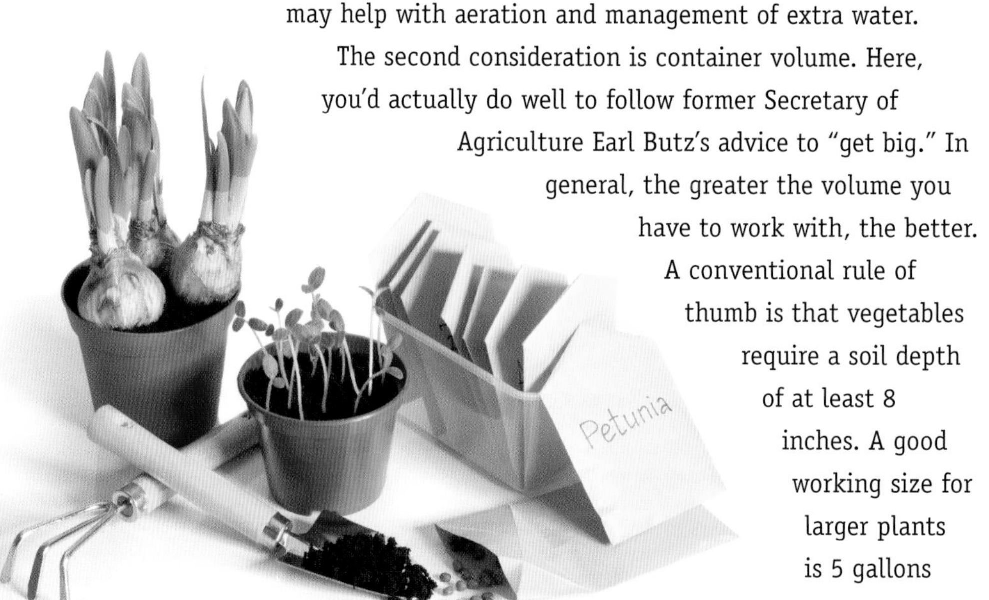

Container Volumes

Some approximate volumes of various containers are:

- 6-inch (15.25-cm) pot: 2.5 quarts (2.4 liters or ⅝ gallon)*
- 8-inch (20-cm) pot: 1 gallon (3.8 liters)*
- 10-inch pot: 2.25 gallons (8.5 liters)*
- 12-inch pot: 3.5 gallons (13.25 liters)*
- 40-pound bag of soil: ≈3.5 gallons (≈13.25 liters)
- ≈35 lb kitty-litter pail: 4 gallons (15 liters)
- Plastic bucket (≈1 foot [30.5 cm] diameter x 15 inches [38 cm] high): 5 gallons (19 liters)

You can calculate the volume in gallons of other boxy containers easily by multiplying (in inches) height by length by width and then dividing the total by 231, since there are 231 cubic inches in a gallon.

- Window boxes/planters:
  - 12 × 6 × 6 inches (30.5 × 15.25 × 15.25 cm): 1.75 gallons
  - 24 × 8 × 8 inches (61 × 20.5 × 20.5 cm): 6.5 gallons
  - 24 × 10 × 10 inches (61 × 25.4 × 25.4 cm): 10 gallons
  - 30 × 8 × 8 inches (76.2 × 20.5 × 20.5 cm): 8 gallons
  - 30 × 10 × 10 inches (76.2 × 25.4 × 25.4 cm): 13 gallons
  - 30 × 12 × 12 inches (76.2 × 30.5 × 30.5 cm): 18 gallons
  - 36 × 8 × 8 inches (91.5 × 20.5 × 20.5 cm): 10 gallons
  - 36 × 10 × 10 inches (91.5 × 25.4 × 25.4 cm): 15.5 gallons
  - 36 × 12 × 12 inches (91.5 × 30.5 × 30.5 cm): 22 gallons
  - 48 × 8 × 8 inches (122 × 20.5 × 20.5 cm): 13 gallons
  - 48 × 10 × 10 inches (122 × 25.4 × 25.4 cm): 20 gallons
  - 48 × 12 × 12 inches (122 × 30.5 × 30.5 cm): 30 gallons
- EarthBox: 15 gallons (57 liters)
- Half whiskey barrel: 20 gallons (76 liters)
- Half wine barrel: 30 gallons (113.5 liters)

*Approximate capacities taken from "Container Vegetable Gardening," publication PM 870B, University Extension, Iowa State University.

These sunflowers are growing above solid cement in less than 4 inches (10 cm) of soil. So what are you waiting for?

(19 liters), and 5-gallon (19-liter) buckets are readily available at home-improvement stores. Vegetables such as tomatoes and peppers prefer the deeper (bucket) variety, while many leafy greens and shallow-rooted vegetables such as small radishes can probably get by in the window box. Shallower containers can accommodate small herbs and some ornamentals.

Don't limit yourself to the rule of thumb, however. You'll never grow a 7-inch (18-cm) carrot in 5 inches (13 cm) of soil, obviously, but many other crops can withstand shallower-than-ideal plantings if regularly watered, well tended, and given vertical support as necessary.

Note that you can buy square pots, which are very convenient for built environments. You'll find nary a right angle in nature, but your living space is

Easy to grow in containers, catnip also repels mosquitoes, cockroaches, and even termites. Then again, it attracts cats. You do the cost-benefit equation.

almost certainly filled with them. One of the advantages of using square containers in squarish environments (balconies, for instance) is that they hold a significantly greater volume of soil—about 27 percent more—than a round container of the same height and diameter. (The percentage decreases if you are using multiple round pots, which can fit together more closely when staggered, but the difference is still significant). Many kitty-litter buckets are rounded squares, so if you're not a cat owner yourself, consider calling on a friend who is. You can grow catnip in exchange, or even some varieties of kiwi vine, which are surprisingly seductive to cats.

The high-end option in the world of containers is the self-watering pot, which is basically a container with a reservoir that holds extra water. When you water a plant in the container, excess water flows into the reservoir. Then, when the soil begins to dry out through surface evaporation or root intake, more water wicks up from the reservoir by capillary action. This kind of container will not eliminate the need to water, but it will stretch out the time between waterings as well as maintain a more consistent moisture level. You can buy self-watering containers or create your own, as described at the end of this section, if you have gumption and a power drill.

The third consideration in container choice is its construction material, which involves both function and aesthetics. Containers can be made of just about anything that can hold a clump of soil in place. Alpine plants, for example, are

Tire Gardens

Old tires are popular garden containers in much of the developing world. They're relatively cheap, portable (they roll, after all), and long-lasting, with a decent capacity for soil. You can even prop them up off the ground using bricks or wood for ease of gardening or to protect crops from small animals, soil-borne pests, or occasional flooding. The big question is whether they are safe to use. According to *Mother Earth News, in the short term, probably yes; in the long term, probably no, thanks to harmful chemicals that may leach into the soil over time.*

But let's say you're committed to taking a tire for a spin during the next growing season. The simplest way to use it is to lay it down on its side and fill it with soil. If you worry about the quality of the ground soil or want the tire garden to be portable, however, you'll want to turn it into a closed container by lining the inside with plastic and—if you want to be fancy about it—cutting off the top sidewall and placing it upside down at the bottom to hold the plastic in place.

traditionally placed in a hollowed-out trough of tufa (a light, porous rock) or one made out of hypertufa, a manufactured alternative. In some parts of the world, root vegetables are grown in hanging sacks. Columns of tires or even single tires are also popular (see sidebar below). For the most part, however, there are four main rigid options—pottery, metal, wood, and plastic—and flexible fabric.

Pottery can be very attractive, but it has a number of significant downsides for the active urban farmer. For one thing, it's very heavy. A 2.5-gallon (9.5-liter) glazed pot might weigh 10 pounds (4.5 kg) empty, and the soil will weigh another 20 pounds (9 kg) or so. Add a plant and water, and that's something you'll rarely want to move. Unglazed pottery, such as terra-cotta, may be a bit lighter, but its porosity can cause significant water loss through evaporation. That can be OK for some dryness-loving houseplants (and herbs), but many vegetables are a thirsty bunch. In addition, pottery is more breakable than you might want for an urban farm that you expect to be heavily trafficked, and ceramic pots are particularly susceptible to cracks in colder zones through freeze/thaw cycles.

From a physical perspective, metals are very suitable for garden containers. They are strong, relatively light, and not porous. However, they can heat up dangerously in the sun and be problematic from a chemical perspective. Aluminum, zinc, and copper can all be toxic to plants. For example, copper looks great, can develop a nice verdigris, and is sometimes used in crop fungicides, but it can also be very dangerous—check out the ingredients of root killer. Brass comprises both zinc and copper, both of which are toxic. Bronze also contains copper, and sometimes aluminum as well. Galvanized steel contains zinc and can be contaminated with cadmium, which may be fine for plants but not for you. According to the Centers for Disease Control and Prevention, high doses of cadmium over a short period "severely irritates the stomach, leading to vomiting and diarrhea," while low doses over a longer term can lead to cadmium buildup, kidney disease, lung damage, and fragile bones.

Whether or not a metal's potential to harm will actually be a problem depends upon a lot of factors. So, unless you're a chemist and know better, I'd recommend avoiding metal, with the possible exception of stainless steel if you can find it.

Wood is relatively tough, light, and cheap—all attractive qualities for containers. It's even organic—in the sense that it contains carbon and has been alive, if not in the sense that it grew without synthetic inputs. The problem is that wood rots. You can buy treated wood or preserve it with chemicals yourself, but then you have to weigh the safety of the additional chemicals in your growing container against the value of prolonging a wooden container's useful life. Half-barrels are often charred to prevent rot, which is preferable to chemical additives if you're planning on growing food.

Whatever kind of wooden container you use, make sure it drains well. Another solution is to line the inside of the wood container with something impermeable, which takes us to the next option: plastics. You can even buy plastic liners for those half-barrels to make a fish pond or long-lasting planter.

Plastics are probably the most flexible of container materials. They can be extremely light, exceptionally strong (think of fiberglass, which is plastic reinforced with glass fibers), and can take just about any shape. Many, but not all, plastics are recycled. If you are going to use plastic, however (and most of us will), consider making two practices part of your life. First, reuse what you can—those kitty-litter buckets being a perfect example. Even though the plastic itself may be recyclable,

Clockwise from the top: Thai basil, butter bean, and Cossack pineapple (*Physalis pruinosa*) grown directly in a bag of soil.

Urban farmers often find creative ways to reuse plastic containers for small-space growing.

reuse eliminates the energy (of manufacture, shipping, and sale) and resources used to make the new container you would otherwise purchase, as well as the resources involved in the recycling process itself. Second, if you do buy plastic containers, try to buy easily recycled kinds, such as polyethylene terephthalate (PET, indicated by the number 1 in the triangle on the bottom of a container) or high-density polyethylene (HDPE, number 2 in the triangle).

The simplest plastic container is a bag of soil. Punch a few drainage holes near the bottom of the sides, cut a slit in the top, and plant directly in the bag. You can buy specialized grow pots to facilitate things, but you don't need them.

Whatever type of plastic you buy, use the container to death. One caveat: avoid reusing plastic containers that have contained potentially harmful chemicals, such as solvents. You can reuse ones that have held salts (halite, for example) after you clean them thoroughly.

Capillary Action

Capillary action is the mechanism that keeps plants hydreated. Water is mildly cohesive, meaning that it sticks together. When water transpires from leaves, it pulls up water molecule below it, which pull up the molecules below them, and so on, like a big chain. Somewhere between 400 and 450 feet, however, the force of gravity overcomes the cohesive power of water, marking the theoretical limit to which a plant could grow. The tallest known specimens of hardwood (mountain ash, Eucalyptus regnans) and softwood (coast redwood, Sequoia sempervirens) seem to confirm the presumed cutoff, both maxing out at around 380 feet tall.

Sorry, I Have Other Plans...

Be aware that there are whole communities of agricultural MacGyvers out there, explaining in detail how they make homemade versions of otherwise store-bought items, many more ambitious than the ones just described. If you try web searches such as "how to make [insert item you want in quotes]," you are bound to find step-by-step instructions, helpful diagrams and pictures, or even videos. FYI, it's also good to search with some commonly used acronyms: DIY for do-it-yourself and SWC for self-watering container, for instance.

The most welcome new addition to container varieties are those made from felt-like fabric. Fabric containers are available in myriad sizes from less than a gallon to 30 gallons (113.5 liters) or more, and are lighter and more easily stored out of season than the other kinds of containers. In addition to being convenient for you, they are well suited to plants. Fabric containers drain well and permit air to reach the roots, encouraging healthy development. They are also far less likely to overheat in summer. While fabric planters might not last as many seasons as some other forms of container, their many advantages recommend them in many circumstances. You can even buy them in square-foot size, allowing you to plan out your garden with ease.

DIY Containers

Plastics are great for creating all kinds of specialized containers, including self-watering and upside-down planters. Self-watering containers have a reservoir of extra water and a means of wicking it up to the soil. Think of a hurricane lamp. The broad wick extends down into the kerosene or other fuel and brings it up by capillary action to the point you light. The wick hardly burns, but the reservoir of fuel steadily empties. So it is with self-watering containers, which hold an extra supply of water and a means of delivering it gradually to a plant growing within. The reservoirs allow you to stretch out waterings and reduce the chance of plants suddenly giving up the ghost on a scorching day. Three popular styles are the hurricane lamp, the recessed hurricane lamp, and the periscope.

project

Hurricane Lamp Style

Potentially holding a great deal of water, the hurricane lamp style suits those looking to maximize the time between waterings. Because it stands taller than the other styles and can be relatively top-heavy, you should ensure that the reservoir does not get too low, which will increases the risk of wind toppling the whole thing.

What You'll Need

- a bucket with a lid to act as a reservoir
- a plant container, smaller than the bucket size, with a drainage hole (or the ability to be punctured)
- a synthetic wick, a few inches longer than the bucket is deep
- a drill or razor

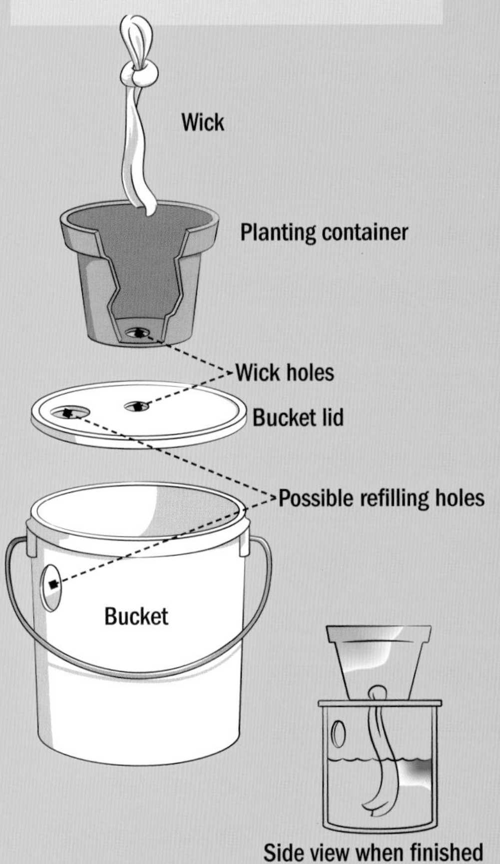

1. If the plant container does not have a drainage hole in the bottom, cut or drill one about half an inch (1.25 cm) in diameter.
2. Drill or cut a small hole in the center of the bucket lid (about 3/8 inch [1 cm] in diameter) and another one on the lid or just below the lid (about 1 inch [2.5 cm] in diameter) to allow for refilling.
3. Tie a loose knot in the wicking material, which can be an old shirt, a dish towel, or a sheet of another synthetic material, such as polyester. (Cotton would work, but it will eventually rot.)
4. String the long end of the wick through the bottom of the plant container, then through the hole in the bucket lid, and attach the lid to the bucket.
5. Fill the bucket with water and any soluble fertilizer you want to use, and fill the bottom of the plant container with soil.
6. Water the soil lightly to make sure it is moist— utterly dry soil will not permit capillary action. Monitor the water level through the refilling hole. It's a good idea to top off when the bucket is roughly half empty, just to maintain enough weight for stability.
7. Plant seeds or transplants.

Recessed Hurricane Lamp Style

project

In this modification of the hurricane lamp method, the plant container is recessed into the reservoir bucket, adding stability and giving a tidier appearance.

1 Before you make any holes, ensure that the plant container rests comfortably on the rim of the bucket, with most of the container dipping down inside the bucket. Only proceed if you have a good fit.
2 Mark on the side of the bucket the approximate depth reached by the plant container.
3 Drill two holes: an overflow hole (about 3/8 inch [1 cm] in diameter) on the side of the bucket just under where you marked, and a refilling hole (about 3/4 inch to 1½ inches [about 2 to 3.75 cm]) on the side of the bucket just above where you marked. (The overflow hole lets you know when you've filled the bucket with enough water and lets excess water out of the bucket to allow air to reach the roots.)
4 If the plant container does not have a drainage hole in the bottom, cut or drill one about ½ inch (1.25 cm) in diameter.
5 As with the original hurricane lamp method, string the long end of the wick through the bottom of the plant container.
6 Place the plant container inside the bucket.
7 Fill the bucket with water and any soluble fertilizer you want to use, and fill the bottom of the plant container with soil.
8 Water the soil lightly to make sure it is moist—utterly dry soil will not permit capillary action. Monitor the water level through the refilling hole. It's a good idea to top off when the bucket is roughly half empty, just to maintain enough weight for stability.
9 Plant seeds or transplants.

What You'll Need

- a bucket without a lid
- a plant container with a lip slightly wider than the bucket top and a drainage hole (or the ability to be punctured)
- a synthetic wick at least 12 inches (30.5 cm) long

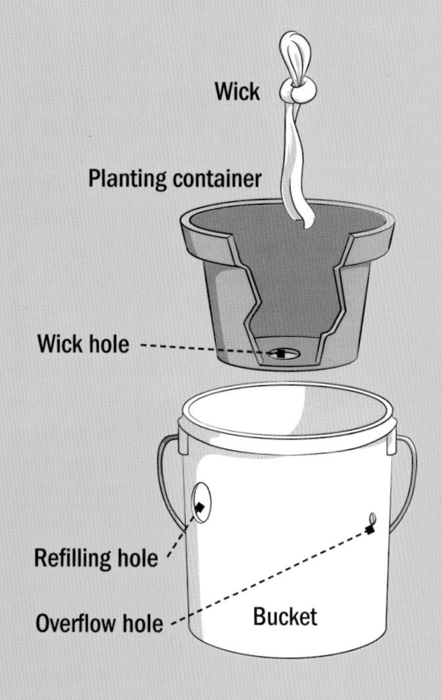

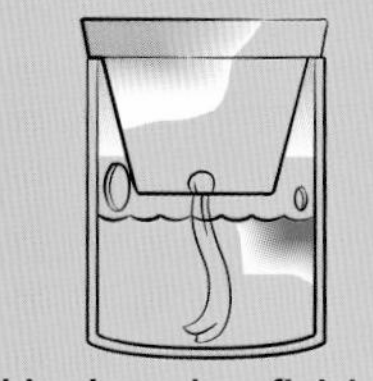

Side view when finished

project

What You'll Need

- two 5-gallon (19-liter) buckets
- a plastic pipe about 1 inch (2.5 cm) in diameter and taller than the buckets
- a synthetic wick at least 12 inches (30.5 cm) long
- a drill with bits of three sizes: small (¼ inch [0.6 cm] or less in diameter), medium (about ⅜ to ⅝ inch [1 to 1.6 cm], and large (the same diameter, or slightly larger, than the diameter of the pipe)

Side view when finished

Periscope Style

The periscope style—or styles, because there seem to be dozens of favorite ways of doing it—is one of the most popular among urban farmers using containers. The type described here is among the easiest, involving no glue, mesh, or smaller containers.

The periscope method is very much like the recessed hurricane lamp method, but the nesting plant container is now a large bucket (the "inner bucket"), and a tube (the "periscope") has been added to facilitate watering.

1. With the large bit, drill a hole just large enough to accommodate the pipe on on the bottom of the inner bucket.
2. With the medium bit, drill a second hole large enough to fit a wick in the center bottom of the inner bucket.
3. With the small bit, drill many smaller holes around the bottom of the inner bucket.
4. Mark on the side of the outer bucket the approximate depth reached by the inner bucket when nested (it will probably be about 3 inches [7.6 cm]).
5. With the small bit, drill two or three holes on the sides of the outer bucket just below where you marked.
6. With the small bit, drill a few small holes on the last 2 inches (5 cm) of one end of the plastic pipe.
7. Knot the wick loosely so that there are about 5 inches (12.7 cm) on either end; pull one end through the center hole at the bottom of the inner bucket.
8. Place the inner bucket inside the outer bucket, and the plastic pipe (holes-side down) into the hole in the inner bucket.

9 Fill the inner bucket with soil, taking care to extend the wick upward, and then fill the outer bucket with water through the tube until water starts to come out of the overflow holes.
10 Water the soil lightly and then plant seeds or transplants.
11 Monitor the water level. You can use a long stick or bamboo pole to measure the reservoir's water level through the periscope, or simply water it until water comes out of the overflow holes.

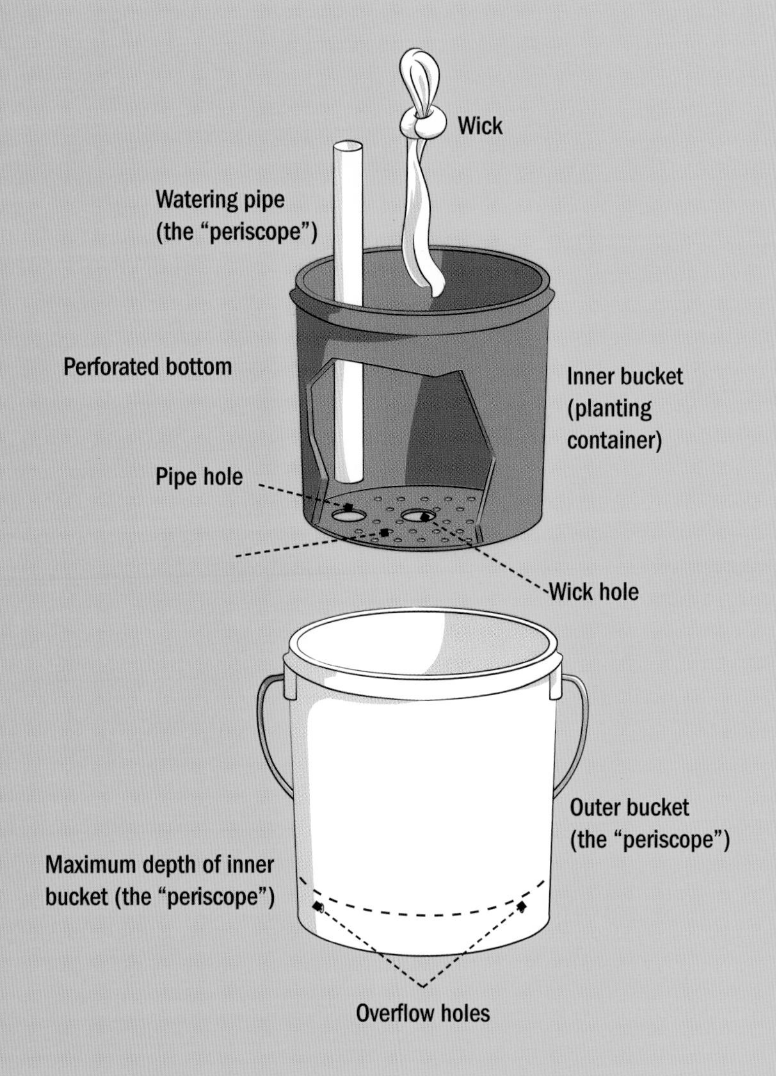

project

Upside-Down Planter

Now that you've mastered these self-watering-container methods, how about trying something more unusual, like an upside-down planter? That's even more easily done.

The difficult part of the upside-down planter is getting the plant inserted into the bottom of the container and held in place—that's where the cardboard, sponge, or landscape cloth comes into play.

What You'll Need

- hanging plastic planter (at least 2 gallons [7.5 liters])
- a drill and medium-size bit (about ⅜ inch to ⅝ inch [1 to 1.6 cm])
- some cardboard, a sponge, or landscape cloth to hold soil in
- a smallish empty container to use for support

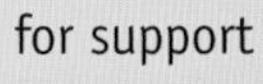

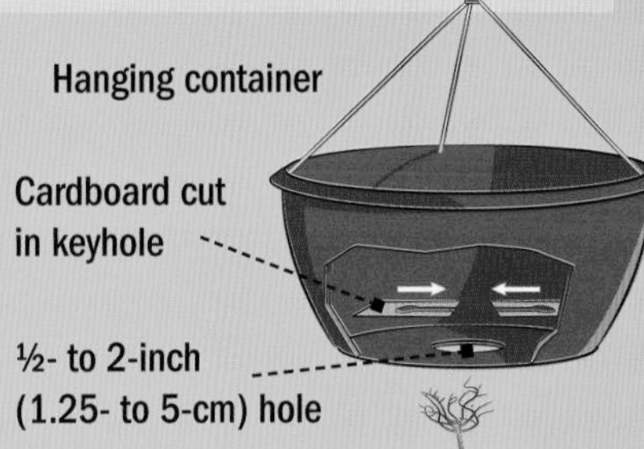

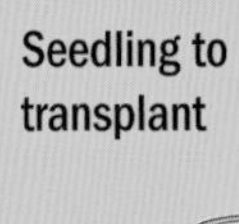

1. Drill a hole in the bottom of the planter about ½ inch (1.25 cm) in diameter if there is no drainage hole that size present.
2. Create an anchor to hold the roots in place:
   - If you're using cardboard, cut two roughly 2-by-2-inch (5-by-5-cm) square pieces, each with a rounded channel (think of a keyhole) cut into one side; the rounded part of the channel should be slightly larger in diameter than your plant's stem.
   - If you're using a sponge or landscape cloth, cut a straight line lengthwise from one side to the middle, where you'll cut a small hole.

Whatever material you use, the point is to hold the plant stem snugly (though not tightly) and prevent soil from falling out of the hole.

3. Take the plant you want to grow and slip the root ball—the mass of soil clinging to the roots—up through the hole at the bottom of the planter. Hold the roots in place with one hand, and then rest the hanging planter on top of the center of the empty container (so that you can rest the hanging planter down without crushing the plant).
4. Once the hanging container is resting and you're holding the root ball with one hand, fit the two cardboard pieces around the stem, facing opposite directions, or open the sponge/landscape cloth along the slit and wrap its center hole around the stem.
5. Fill in enough potting mix to completely bury the root ball—at which point the plant will be securely attached—and then fill up the rest of the container with potting mix. Voila!

Note: Hanging planters tend to have a low volume of soil and a large surface area for evaporation. Check the soil moisture level frequently, and water as needed.

That Old Standby: The Ground

The most time-tested method of farming is to simply grow plants in the ground: put in some seeds, add water, watch them grow. There's plenty of land to make this happen in most cities. A 2000 report, for example, noted about 900 acres of vacant land in Trenton, New Jersey; 1,200 acres in St. Louis; and 2,500 acres in Milwaukee. Even if the land is there, however, you may not be able to access it. If you can, you may still need to face the three big obstacles to in-ground urban plots: soil compaction, poor fertility, and contamination.

If compaction is the only problem, you can work around it. You can break up the soil with a rototiller, employ a double-digging technique, shovel or fork in organic matter, and/or grow plants that break up the soil, such as sunflowers, daikon radish, or alfalfa. You can also put a raised bed that's high in organic matter directly above the compacted soil to use as another growing medium during the time it takes the soil to break up through the action of roots and soil life.

All of these methods can also help restore soil fertility. It may seem contradictory to say that cities tend to be built above the most fertile land but then cite poor fertility as a common problem of urban soil. The disparity is explained by human activities. Excavation destroys the natural layering of soil at dig sites while also compacting the ground underneath the excavation equipment. Predevelopment agricultural activity may have depleted soil nutrients without replacing them at a particular site, while industrial activities can leave behind a toxic legacy. The most immediate solution is incorporation of organic matter and fertilizers, though certain plants will also help rebuild fertility over time. Cover crops, for example, are grown to help restore the soil, prevent erosion, and suppress weed growth in between other plantings. Many are legumes such as peas, which "fix" atmospheric nitrogen into the soil, or deep-rooted plants such as alfalfa, which break up the ground and draw up nutrients from below.

Contamination is, by far, the biggest concern. You may not be able to see it, smell it, or taste it in crops grown there, but it can be there—even on the White House lawn. Lead poses one of the biggest risks in older cities because of decades of both lead-paint usage and the use of leaded gasoline by a relatively heavy concentration of traffic. Unfortunately, lead is just one of a long list of contaminants.

Plants can be fantastic accumulators of specific elements, which is great when spinach concentrates iron from the soil for you, or oranges bring up calcium, but is not so great in other instances. For example, a wildflower called Alpine pennycress is so good at accumulating zinc from high-zinc soil that its ashes contain about the same percentage of zinc as high-grade mined ore. Broccoli can take up dangerous levels of selenium from the soil, and sunflowers can accumulate arsenic, lead, and even uranium. Suffice it to say, you don't want to eat vegetables rich in these elements. This doesn't mean you shouldn't grow and eat broccoli and sunflower seeds. It does mean, however, that you shouldn't do so in soil contaminated with those specific elements.

How do you know if the soil is contaminated? Soil tests can check for specific contaminants, such as lead, but a standard test will not identify every possible contaminant. The first thing you should do is perform a little due diligence.

First, just look around you. Is the site next to a busy roadway or an old painted building? If so, lead could be an issue. If it is next to a building, is the building made of wood or other materials? An old wood-framed building might have been ringed with chlordane or another termite-control chemical. Is the site littered with potentially toxic materials, such as rusting cars, cans of solvent or other chemicals, or baited rodent traps? Those could signal problems, too.

Next, find out what you can about the history of the site you want to farm. A former body shop is much more likely to leave legacy contaminants, for example, than a former cupcake factory. You might want to contact the site owner or see if you can trace the ownership at the city clerk's office.

Once you've taken in all the site observations and historical data you can (or are willing to), consult your state's cooperative extension and your city's environmental department to see about soil testing. Some will perform free tests for community gardens or farms, so it is worth contacting both. They may also know if there are particular contaminants common in your state or city. Describe your site and ask what they suggest testing for. If they do not test for a particular contaminant that concerns you, ask for a recommendation to a commercial lab. In all likelihood, you will have to pay for a test, but it is money well spent.

My soil test, for example, revealed that my soil was brimming with potassium, moderate in phosphorous, and low in nitrogen. So I didn't need to add any potassium (which is included in every "balanced" fertilizer) to the soil, and the soil test report recommended specific amounts and forms of nitrogen and phosphorous to add. The report also told me that my soil included 6.2 percent organic matter and sufficient micronutrients.

I also learned that the soil possessed a "medium" range of lead. That's not unusual for urban areas, but don't let that trick you into swearing off urban produce; an apple you pick yourself at a country orchard may well be more contaminated than one you grow in the city lot next door. In fact, rural communities are some of the hardest hit by pollutants. Arsenic, lead, and even mercury have been used historically as pesticides in agricultural areas, not to mention organic compounds such as DDT.

We live in a polluted world. That's life. Being aware of the risks is the best way to minimize them, and there are many practices to help the urban farmer remain safe. A soil test can help greatly. Even if it reveals the presence of something, such as heavy metals, it also reveals soil pH, a key factor in the bioavailability of many contaminants. The test can guide you into adding organic matter or lime, both of which can help neutralize the effects of many contaminants. Adding mulch, in addition to organic matter, will help prevent contaminated dust from being kicked up or splashing up on crops when it rains. Similarly, washing leafy greens and peeling root crops after harvest greatly reduces any lead contamination.

Let's assume you're stuck with a mildly contaminated site. If there has been lead paint used or termite controls applied, the heaviest concentrations will be around the foundation of the building. If these concern you, choose ornamental plants for such areas. In other areas, you might do well to follow the suggestions of the Cornell Waste Management Institute (CWMI), which lists on its website (cwmi.css.cornell.edu) several of the recommended crops for sites known to be contaminated with heavy metals as well as the crops to avoid entirely. Among the best are tree fruits, tomatoes, peppers, cucumbers, and peas. Among the ones to avoid are leafy green vegetables, broccoli, cauliflower, snow peas, and root crops. (You can find the URL for CWMI's publication, "Soil Contaminants and Best Practices for Healthy Gardens," as well as related resources, in the Notes & Resources section).

Finally, if contamination concerns are too great for you to plant directly in the ground, consider building raised beds. Build the beds over landscape cloth (which prevents roots from penetrating the contaminated soil), and you can take advantage of the site's cultural benefits (light and water) without suffering from the soil's problems.

Raised Beds: Better than the Ground?

Raised beds are simply areas of cultivation built upward from a surface. The soil is usually (though not always) held in place by a wooden frame, cinderblocks, or something similar. In addition to raised beds' providing a safe growing medium above contaminated ground, they offer many other advantages by providing:

- better drainage;
- workable soil above hardpan, rocky ground, or even asphalt;
- a longer growing season because they warm earlier in spring and cool later in autumn;
- a convenient base for installing row covers (gauzy textile used to protect plants from pests), a hoop-house frame, or fencing; and
- an elevated surface that requires less stooping.

Raised beds that can be accessed on all sides are typically around 4 by 8 feet (1.25 by 2.5 meters). There's also a certain beauty of economy in being able to purchase three pieces of nominal lumber (let's say 2 inches by 8 inches by 8 feet [5 cm by 20 cm by 2.5 m]) and cutting just one of the boards in half. Then you have two 4-foot (1.25-m) boards and two 8-foot (2.5-m) boards, all ready to be assembled into a rectangular raised bed. You could also buy just two 8-foot (2.5-m) boards and then saw each into a 5-foot and a 3-foot section to make a rectangular 3-by-5-foot (91.5-cm-by-1.5-m) raised bed.

It's thought that 2 to 3 feet (61 to 91.5 cm) is about as far as noncontortionists can lean over a bed without crushing something, so the short dimension should be no longer than a yard if it's against a wall or you can otherwise access it from only one side. Also keep in mind that wood will begin to bow beyond 8 feet (2.5 m) or so, so you would need to brace longer beds.

Raised beds made of wood can be nailed or screwed together, although using a few L-brackets for reinforcement is a good idea. Using sections of 4-by-4-inch (10-by-10-cm) post to mark the corners (and possibly anchor the beds into the ground) makes it easier to ensure that you have perfectly right angles. You can also buy all manner of raised-bed-securing corners and joiners from gardening catalogs. If you're going to do it yourself, I'd recommend securing the corners using large hinges. This allows you to form the box into an angular shape that might better fit your space and even to fold up some sections for storage without needing to disassemble the whole thing.

Raised beds can prove as beautiful as they are practical.

There's some disagreement about whether or not the chemicals in pressure-treated wood (which used to commonly include arsenic) can leach out in quantities enough to harm the eaters of food grown nearby, but I'd err on the side of caution. Ditto for wood treated with creosote. You won't want to be the victim of irony, building raised beds to avoid contamination only to contaminate the new soil with the materials used for your raised bed.

The downside of untreated wood, of course, is that it tends to rot sooner. All wood eventually rots, but you can take some steps to maximize the life of a wooden raised-bed frame. First, choose the right wood. If you're buying commercial lumber, choose kiln-dried (but untreated) cedar, cypress, or redwood, all of which will last one or more years, depending on your climate, soil life, and other conditions. A number of sources recommend charring the side of the wood that will face the inside of the raised bed—if you're one of those city dwellers who happens to have a blowtorch handy. This will not prevent rot, but it may slow it down.

If your ground is clean and you want to plant directly above it without a barrier, at least cover any lawn with several layers of newspaper or biodegradable landscape cloth. Both will decompose readily yet help minimize the grass's invasion of your beds. Figure on around 4 inches (10 cm) of new soil as the minimum depth for

Know Your Cloths

Landscape cloth (also called landscape fabric or weed barrier) is a strong, tightly woven material often used in gardening. Its main virtue is that it lets air and water through while physically blocking the development of unwanted plants from below and sometimes blocking the roots of desirable plants from above, as when it is spread over contaminated ground beneath a raised bed with fresh soil. Biodegradable versions have more of a plastic feel and should dissolve within a season or so.

A primary use of landscape cloth is to suppress grass if you're building a raised bed directly on top of lawn. The regular landscape cloth will probably suppress the grass better—and certainly for longer—but it will also prevent the roots of your crops from reaching below the level of the raised bed. Biodegradable landscape cloth will not do quite as good a job suppressing grass, but it will give you a head start, allow roots to head down deep into the soil, and provide a bit of fertilizer.

Hardware cloth is something entirely different: wire mesh. In urban farming, it is employed primarily for construction that needs to let airin and keep pests out, such as in animal pens or compost bins.

planting above lawn or ground, or 6 inches (15.25 cm) of new soil if you're using a landscape-cloth barrier because of contamination. Deeper beds are better but will require stronger frames.

Whatever size bed you are considering, be aware that they hold a lot of soil—a lot. A 4-by-8-foot raised bed with a depth of 1 foot (30.5 cm), for example, holds 32 cubic feet (0.9 cubic m) of soil. That doesn't sound like much, until you realize that it's about a ton of dirt, or about fifty 40-pound (18-kg) bags. If you have a lot of organic matter on hand, such as leaves or grass clippings, you might want to rake those in first and then cover with garden soil. It will reduce the amount of soil you need to buy and enrich it with organic matter. You can then build the bed up over time—more leaves, more soil.

One trick used by raised-bed gardeners is to convert the bed into a hoop house, which can extend the growing season significantly on both ends. Essentially a

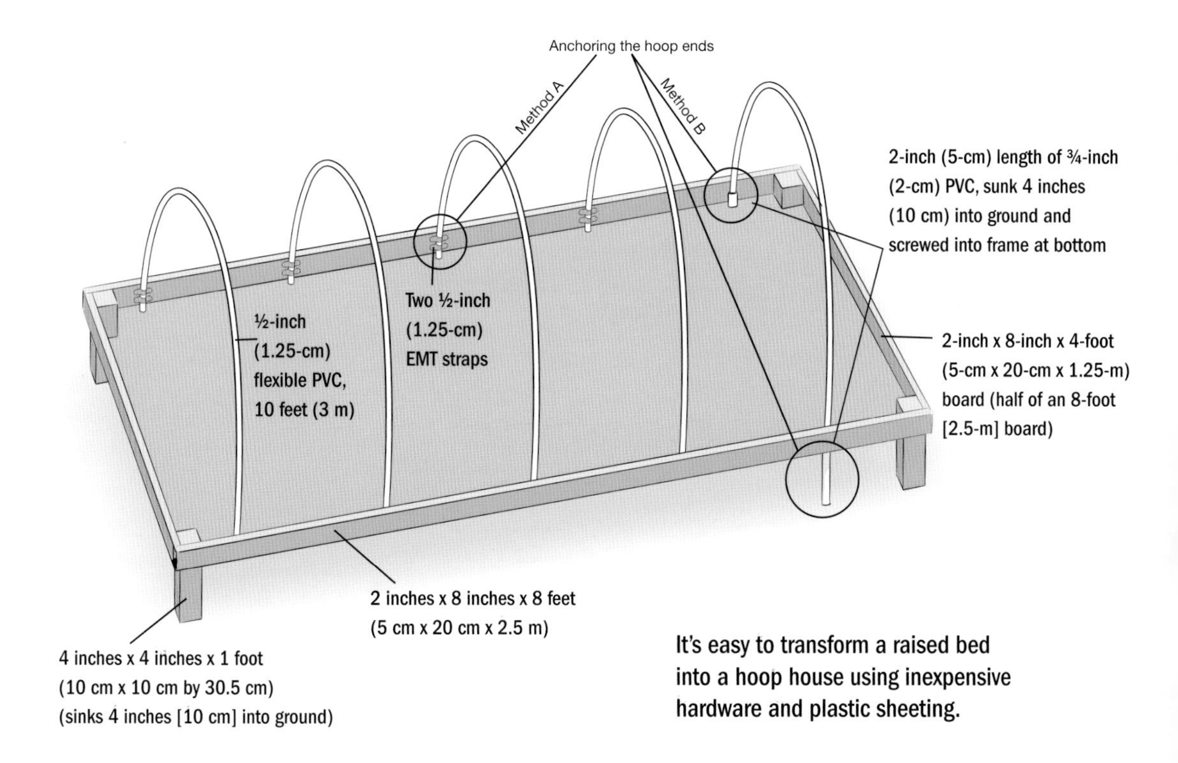

It's easy to transform a raised bed into a hoop house using inexpensive hardware and plastic sheeting.

greenhouse "lite," a hoop house can also be used to accommodate shade cloth (a mesh fabric used to keep some crops cool or protected from direct sunlight) or row covers to protect crops from pests. The simplest method to convert a small raised bed into a hoop house is to arc two or three lengths of flexible ½-inch (1.25-cm) PVC pipe across the narrow dimension of the bed and then cover the PVC frame with plastic sheeting. If you think that a hoop house might be in your future, consider affixing short lengths of slightly wider PVC piping to the inside of your raised bed to anchor the ends of the hoops (like a pen fitting into its cap) or creating braces using electrical metallic tubing (EMT) straps. The flexible PVC tubing is sold in 10-foot (3-m) sections. If you anchor such a section so that it bends over the 4-foot (1.25-m) width of the raised bed, it will create an arc about 4 feet (1.25 m) high.

You can find endless hoop-house variations online that are built on this concept. If you're looking for something a little more substantial, the Samuel Roberts Noble Foundation has developed a portable but sturdy 14-foot (4.25-m) wide high hoop house that can be built from 36 to 80 feet (11 to 24.5 m) long (see Notes & Resources).

Greenhouses and hoop houses offer many advantages, including a longer growing season and protection from pests.

Wick Gardens for Rooftops

Rooftops are not vehicles in themselves the way containers or raised beds are—after all, an extensive green roof is essentially a very large container or extremely raised bed. Both containers and raised beds can be used on roofs, but rooftop gardens also lend themselves to other vehicles, such as wick gardens. ECHO has been developing wick gardens for years as a means to capitalize on flat concrete roofs and available resources in the developing world. As with self-watering containers, wick gardens benefit from capillary action. In fact, a wick garden is almost like a self-watering container turned inside out. In developing countries, people might use old towels or rags as wicking material; you can try those or purchase a specialized membrane, such as capillary matting used in the greenhouse trade.

project

Wick Garden

This is almost certainly the easiest specialized container to make.

1. Cut out an expanse of plastic sheeting slightly larger than your wicking membrane and put it on the growing surface.
2. Lay the wicking material on top of the plastic sheeting.
3. Drill a ⅜-inch (1-cm) hole in the cover of the bucket. Fill the bucket with water and water-soluble fertilizer and invert it (lid side down) onto a corner of the wicking material. The water should begin traveling throughout the wicking material. If you have a thin wick that does not seem to be providing enough water after an hour or so, try doubling it over.
4. Put your seeds or transplants directly on the wicking material. Cover the wicking material with something to prevent water loss through wind and sun exposure. Soil could work, as could rocks, chipped wood, or even crushed soda cans. In addition to preventing water loss, this mulching material acts as a kind of support for the growing plants.
5. Refill the bucket as necessary, and use more than one bucket for larger areas. ECHO suggests using one bucket for every 16 square feet (1.5 sq m).

What You'll Need

- one or more buckets with lids, preferably 3 or more gallons (about 11 or more liters) each
- an expanse of polyester or other synthetic wicking material the size of the area you want to cultivate
- plastic sheeting
- a drill
- mulching material, or landscape cloth to hold soil in
- a smallish empty container to use for support

Next Steps

Once you know your farmstead(s) and vehicle(s) for farming, resist the urge to order huge quantities of seeds or plants. First, ask other urban farmers (if they exist in your city) what has worked and not worked for them. Get to know your local nursery staff. Most of all, perhaps, check out your local cooperative extension, which exists to help people like you. It's not good enough to know that local farmers have had success with tomatoes—find out which tomatoes. Not all tomatoes are created alike. Different cultivars require different growing seasons, prefer different cultural conditions, and resist different pests and diseases. Chances are, the best varieties for your area are probably much, much better than the worst ones. If you choose the best kinds, you're ahead of the game even before you've planted the seeds.

Record all of these bits of information and sketch out the area you have to work with. You can't plant everything you want, so take a close look to consider your own microclimates and conditions. You don't want to plant corn in a wind tunnel, for example, even if it gets a lot of sun. Read up on the plants you want to whittle down the choices. Some plants like morning sun and afternoon shade, for example, and others prefer morning shade and afternoon sun. Knowing the plants' sun preferences will tell you which ones can go on the eastern edge and which on the western edge. And if you have only one exposure, you might need to rule out a slew of candidates. Gauge which areas will get the most rain or water from gutters, and plant accordingly. If you're planting both perennials (such as fruit trees, berries, rhubarb, or asparagus) and annuals (most commonly vegetables, such as broccoli), it often makes sense to place the perennials along the periphery of the site, out of the way of foot traffic. The central area is usually well suited for annual crops.

Consider where you have water access. All other things being equal, it probably makes sense to plant the thirstiest plants (such as tomatoes) near the water source and the most dry-tolerant ones farther away. Think about where you might want to keep a composting bin (see chapter 5 for more about composting). Instead of hiding it away in a dark corner, for example, you might want to give it some sunny real estate, both to accelerate the composting and—if it's filled to capacity and not too hot—possibly to provide a good habitat for vegetables, such as squashes.

Keep track of what you plant each year and how it turns out. You will make mistakes every year, but documenting them will help you make fewer the next year.

Plant appropriately for your climate and cultural conditions. The beautiful Zapotec Pink Ribbed Tomato—here unripe—may thrive in its Central Mexican homeland, but it has performed poorly in New Jersey trials.

Consider the tools you need before you buy anything. All farmers need a trowel (small hand shovel) and pruner of some kind.

Commonly recommended tools for small in-ground farms include a stirrup (or loop) hoe, a collinear hoe, a seeder, and a rototiller. You should be able to buy the first three for less than $200 total, and a used rototiller will cost several hundred (or more) dollars. If there are other farmers in your area, see if there is a local tool library, as there are in some cities nationwide. If there isn't, see if you can get a local nonprofit or city agency to start one.

Once you've decided which kind of farmer you are, where you'll farm, and how you'll farm, consider your growing medium, or substrate. Usually, it's soil. More ambitious farmers might experiment with soilless media based on vermiculite or perlite, or aquaponic setups that grow crops in water.

CHAPTER 5

The Dirt on Dirt

(and Other Substrates)

Plants grow in soil. What could be simpler? Yet soil is not the only surface in which plants live. Some will grow in water, while epiphytic bromeliads and orchids often grow in little more than air. Still, for most of us, most of the time, it's all about dirt. Soil does far more than simply anchor plants, and it is much more complex than most people realize. First, there are the sheer physical dimensions of the stuff. About 90 percent of soil solids consist of rock. The weathered rock that forms the backbone of soils can include relatively large particles of sand (.05–2 mm in diameter), medium-sized silt (.002–.05 mm), and tiny bits of clay (.002 mm or less—the largest ones are less than one-third the diameter of a red blood cell).

Clay, silt, and sand are each considered soil *separates*, which is the term used for particles within given size ranges. Soil scientists recognize twelve major soil types based on these particle sizes and their combinations. Put the right separates together and—voila!—you have the perfect ensemble, known as *loam* among soil scientists. It's the sweet spot right near the center of the soil triangle, composed of roughly 40 percent sand, 40 percent silt, and 20 percent clay.

Soil Textural Triangle

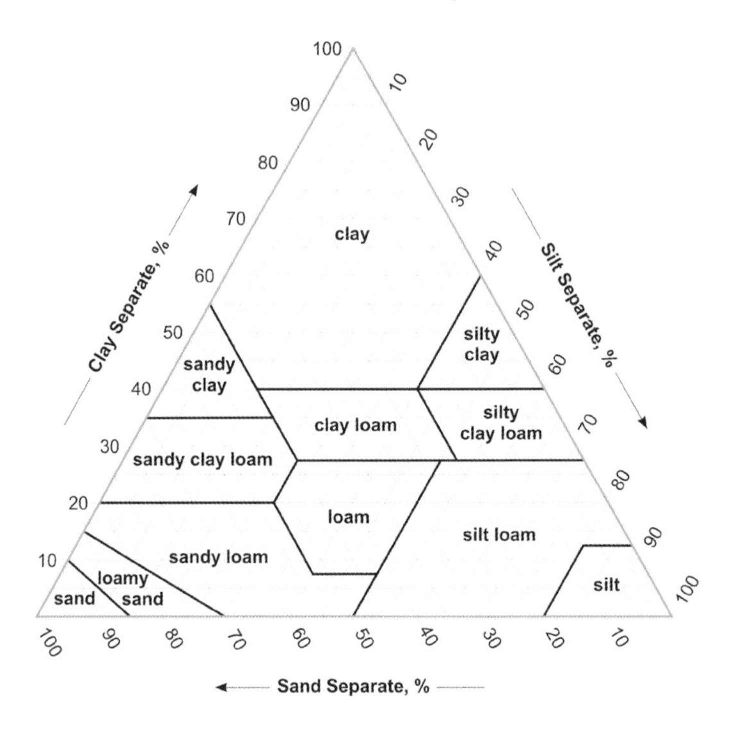

The parent rock of a given soil can also be composed of a huge range of minerals, from the basalt that comprises the black soils in the northwest of India's Deccan Plateau region to the gypsum that gives America's Four Corners its characteristic white soil. The minerals, in turn, provide varying degrees of the elements required for plant growth, including the primary macronutrients (nitrogen, phosphorus, and potassium), the secondary macronutrients (calcium, magnesium, and sulfur), and about half a dozen micronutrients (such as boron).

Just 10 percent of soil comprises things such as organic matter. From a farming perspective, however, it's this tip of the iceberg that makes the real difference. Good soil is often described as resembling crumbly chocolate cake. Just like in a cake, about half the volume of good soil consists of pores that accommodate water and air. These pores are critical for the capillary action that draws water up from the soil depths. In fact, a soil with a healthy structure can hold nearly twice as much water as a soil with identical texture (composed of the same stuff) but poor structure.

Finding the Dirt on Your Soil

The USDA has many online resources to help you figure out your soil type. Loamy sand or sandy loam? If you're able to tease out the separates in your soil and estimate their percentages, you can identify where it falls in the soil-texture triangle. If you're not sure that you can determine the ratio of soil separates by sight, but you are comfortable handling the stuff, the USDA offers a flowchart to help you determine soil texture by feel at www.nrcs.usda.gov/wps/portal/nrcs/detail/soils/edu/?cid=nrcs142p2_054311.

If you're handy with online tools, you can enter your street address and find a wealth of information on your local soils (including separates, precipitation, and landforms) at https://websoilsurvey.sc.egov.usda.gov/App/WebSoilSurvey.aspx. If you just want to know your "official" state soil, you can find it at https://www.nrcs.usda.gov/wps/portal/nrcs/detail/soils/edu/?cid=stelprdb1236841.

Note that your soil may not match the official state soil. You may live in a part of Maine that doesn't have the legislatively honored Chesuncook soil, for example, or a part of Idaho without the volcanic-ash-rich Threebear soil that is officially recognized.

FYI: A Phd in NPK and Ph

Properly applied, fertilizers replace nutrients missing due to agricultural depletion or the native composition of the parent rocks. This usually involves the primary macronutrients, indicated on fertilizer labels by N (for nitrogen), P (for phosphorus), and K (for potassium/potash). The three numbers on a fertilizer label represent the available percentage (by weight) of the macronutrients, always in N-P-K order. A fertilizer listed as 5-12-20, for example, would be (by weight) 5 percent available nitrogen, 12 percent available phosphorus, and 20 percent available potassium.

One of the other soil factors over which farmers can exercise considerable control is a soil's pH, a measure of how acidic or basic (alkaline) the soil is. It uses a 0 to 14 scale, where 7 is neutral—neither acidic nor basic. The lower the number from 0 to 6.9, the stronger the acid. Hydrochloric acid, for example, is about a 2. The higher the number from 7.1 to 14, the stronger the base. Bleach and lye are both strong bases with a pH of about 13. Most plants prefer a pH hovering around neutral (6.5–7.5), though some plants (such as wetland natives) enjoy more acidic conditions, and others (such as alpine and desert natives) tend to like more basic soil.

Organic matter helps create the "glue" that holds soil separates together into helpful little clumps that facilitate pore space.

This granular, cake-like quality (confusingly called *texture* in baking but *structure* in soil science) is critical. Rock particles alone do not make good soil. In fact, it's undigested bits of carbon from organic matter that give most healthy soils their characteristic brown color. Yet even if you could create your own perfect soil blend of 90 percent rock particles—in the ideal 40 percent sand, 40 percent silt, 20 percent clay ratio—and 10 percent organic matter, you still would not necessarily have good soil. That would be like dumping cake ingredients in a bowl and then baking it without mixing. It's not just the ingredients that count, it's how they interact. In soil, the key determiner of how things interact is its living component. Nothing you grow above the soil line will ever compete with the diversity below it. The top 6 inches of an acre of healthy soil can contain several tons of organisms and once-living materials.

A single gram of topsoil (about one-quarter of a teaspoon [1.25 ml]) can be home to a billion bacteria of a million different species. (By comparison, there are just over 5,000 species of mammal on the entire planet.) Also within that 1 gram you can count 100 million fungi, maybe 50,000 algal cells, 1 million protozoa, and 100 tiny worms called *nematodes*. Sometimes plants engage in symbiotic relationships with certain fungi or bacteria. Mycorrhizal fungi, for example, grow like very thin hairs off of many plants' roots. These roots increase the surface area from which plants can take up nutrients, particularly phosphorus. Similarly, *Rhizobia* bacteria infect some plants—specifically legumes—and convert atmospheric nitrogen into a form usable by the plant. In both cases, the plant feeds the microbes carbohydrates and other goodies.

Worms, termites, ants, wood lice, and armies of other soil macrofauna serve as the initial demolition crews, eking out their lives from the dead biotic matter. Fungi pitch in, too. Bacteria are the competitive eaters of all things biotic, breaking down waste and freeing up minerals in the soil for plants. The wonderful earthy smell of good soil mainly results from the spores of actinomycetes, stringy (or filamentous) bacteria that resemble fungi. All of this activity results in gums, starches, undigested waxes, and other bits that help stick particles together in helpful little clumps—the crumbs of the metaphorical chocolate cake. Plants play an important role, too, physically breaking up soils, sloughing off cells, bringing up nutrients, and ultimately dying and contributing more organic matter for the buffet. When all of the dead biotic material is sufficiently decomposed to a point of stability, it becomes *humus*—the magical stuff of soils.

The Pharmacy You Walk On

In addition to its many other benefits, compost is credited with suppressing organisms that cause disease in plants. After all, the bad bug and the good bug have been occupying the same patch of soil for who knows how long, so the good bug knows all of its enemy's moves. Good bugs even produce antibiotics to suppress the bad ones. In fact, many of the antibiotics we commonly use, such as streptomycin, have been isolated from soil. Vancomycin hails from Borneo, Fujimycin from Japan, oritavancin and daptomycin from Haiti, platensimycin from South Africa, and chloramphenicol from Venezuela. The immunosuppressive drugs cyclosporin and rapamycin come, respectively, from soils in Norway and on Easter Island. The soil fungus *Penicillium chrysogenum* feasts on organic matter in the soil, from which it also colonizes vegetables and fruits by airborne spores. The *P. chrysogenum* strain responsible for mass production of penicillin came from a moldy cantaloupe in Peoria, Illinois.

With all of these bugs and the antibiotics excreted by humans and livestock into the environment, you'd think that the ground beneath us would be swimming in antibiotics. Scientists have recently discovered, however, that there are other bacteria that gobble up antibiotics like bowls full of popcorn. The good news is that these bacteria could be helpful in cleaning up waste contaminated with antibiotics before they induce resistance among human pathogens such as *Staphylococcus aureus*. The bad news is that there's a risk that human pathogens currently treatable with antibiotics will develop the antibiotic-munching ability.

Soil Amendments and Conditioners

If soil is such a wonder, you might ask, why is there a subtitle called "Soil Amendments and Conditioners"? The reason is that human activities, particularly agriculture, can deplete soil to the point of its being only marginally useful. Farming can also be done in such ways as to enrich the soil—one of the goals of organic agriculture—but probably at a greater cost of time and money. Unfortunately, cheap usually wins, even when people should know better. *The Rodale Book of Composting*, which may be the subject's bible, recounts how Thomas Jefferson wrote to George Washington that it was simply cheaper for him to buy new land than to fertilize his existing land with manure—until it became more expensive. Faced with land depleted by cash crop after cash crop, Jefferson eventually began experimenting with manure and compost, a method to which we may all be soon returning.

Soil can take so long to restore that it is sometimes thought of as the least renewable resource. Jefferson's experience played itself out again, and on a much worse scale, in the Great Plains. Before it was put under the plough, this area was the playground of bison. Assuming 50 million bison, producing 6.5 tons of dung per animal each year, and a 500,000-square-mile (almost 1.3-million-sq-km) range, that's about .42 pounds (0.2 kg) per square yard (0.8 sq m) each year, a mere sprinkling. Over the four centuries before the Western land rushes of the 1890s, however, we can estimate the production of about 168 pounds (76 kg) per square yard (0.8 sq m), which, if not decomposed, could blanket the Great Plains in roughly 8 inches (20 cm) of high-quality buffalo pies. Yet it would be decomposed, of course, by uncountable tons of worms and bacteria, which would have also feasted on deep-rooted grasses and the leavings of prairie dogs, birds, snakes, ants, and a range of biota, together dwarfing the biomass of bison. In other words, it was a well-tended organic garden of extreme richness. Within forty years of traditional agriculture, a drought turned it into a dustbowl.

Just imagine what a century or more of cultivation has done to our agricultural lands. Thankfully, never have so many options for soil amendment been at our disposal. These options fall into two big categories: inorganic and organic. To avoid confusion between "organic" in the commercial sense (grown in a certain way) and in the chemical sense (containing carbon and once alive), let's call inorganic "abiotic" and organic "biotic."

You can approach the determination of what to apply (if you need to) from an artsy standpoint—if water puddles on the surface of your soil, for example, add something to improve porosity and drainage, such as vermiculite or expanded clay pellets. If you have sandy soil that endlessly drinks up water, add biotic matter. Whether sandy or clay, acidic or basic, soil almost always benefits from the addition of compost. If you have Internet access and a knack for plant identification, you can search for "indicator weeds" to see if local weeds can give you a clue to the composition of the soil (clover might indicate nitrogen-poor soil, for example, since it can fix its own nitrogen from the atmosphere, unlike some other weeds). A more scientific approach is to get a soil test, which will give you much more specific information and often, as a bonus, specific advice on what to add to the soil.

Abiotic

While biotic amendments all affect the soil chemically, abiotic amendments can be nonnutritive—essentially physical—conditioners, such as polystyrene pellets, or nutritive, such as Epsom salts.

Nonnutritive

The urban farmer has a range of abiotic matter available to amend soils. Many can be used as sterile growing media by themselves. Such soilless media has several advantages. They are typically sterile because of being processed at very high temperatures. While a lifeless growing medium is anathema to soil lovers, it is convenient if you fear soil-borne pests. Further, many abiotic media can be reused.

More frequently, however, these nonnutritive abiotic materials are blended into potting mixes to improve porosity, lighten the weight, and help with moisture management. Three of the main four nonnutritive abiotic soil amendments (perlite, vermiculite, and expanded clay pellets) are like popcorn, i.e., puffed up under tremendous heat. The fourth (rock wool) is very much like cotton candy.

Perlite: Perlite looks suspiciously like Styrofoam crumbs (as a matter of fact, polystyrene is sometimes used—fraudulently or just cheaply—as a substitute). Perlite is actually a glass that has been expanded using heat to about ten times its natural volume. It is extremely light (about one-tenth the weight of an equal volume of typical soil) and can hold three to ten times its weight in water. Perlite

Sterilizing Soil

Healthy soil contains far more beneficial critters than harmful ones, and compost-enriched soil has been shown to fight damping off (a fungal disease that shrivels young plants) and other problems. However, if you are of the bleach-is-next-to-godliness school of gardening—or have identified pests, such as fungus gnats, in your soil—sterilization may be the way to go.

You can sterilize your own soil in a variety of ways. To do it in a conventional oven, set the temperature somewhere in the range of 210 to 250 degrees Fahrenheit (99 to 121 degrees Celsius), put the moistened soil in an oven-safe container, and cover it with aluminum foil. Use a cooking thermometer to make sure that the center of the pile hits at least 180 degrees Fahrenheit (82 degrees Celsius), and then cook for thirty minutes. You want to be sure that the soil is heated to at least 180 degrees Fahrenheit (82 degrees Celsius) but does not exceed 200 degrees Fahrenheit (93 degrees Celsius), for thirty minutes. Adjust the oven temperature as necessary to keep the soil in the 180-to-200-degree-Fahrenheit (82-to-93-degree-Celsius) range. Keeping an oven-safe thermometer probe in the soil can help.

To sterilize soil in a microwave oven, put about a pint of moistened soil in a microwave-safe plastic bag (left open) and microwave it on high for about two and a half minutes. If you're uncertain about the exact soil volume or power of your microwave, you can always let the soil cool for about ten minutes and then blast it again. Just remember that you cannot put metal in the microwave—no aluminum foil, rusty nails, or the like in the soil.

doesn't actually absorb water—its pores are closed—but it holds water along its surface area; fine perlite holds more than coarse perlite. Perlite tends to be slightly alkaline (pH 7.0 to 7.5) and resists compression. If a container is heavily drenched, perlite will often float to the top.

Polystyrene: Polystyrene foam is actually quite similar to perlite in performance, with less potential for holding water on its surface. The main downsides are that it is a seldom-recycled plastic, that static electricity makes it stick all over the place during handling (as we've all experienced with packing peanuts), and that it apparently shrinks during steam sterilization, which would likely affect commercial horticultural operations more than most urban farmers (except for those with access to steam sterilization equipment). Like perlite, polystyrene is apt to float to the top of a heavily drenched container.

Vermiculite: Vermiculite is mica—a flaky sheeted mineral—transformed in much the same way as perlite to about ten to thirty times its original volume. It is about the same size as perlite, but shinier and brownish. It is also very light, can hold about six times its weight in water, and is slightly alkaline (pH 7.0 to 7.5, sometimes higher). While mainly nonnutritive, it does provide trace amounts of potassium, calcium, and magnesium. Unlike perlite, which resists compression, it can be compacted down permanently in volume. This is one reason it is used more frequently in seed starting (where compaction is not likely to be a problem) than in higher-volume plantings.

Expanded clay pellets: Extreme heat allows manufacturers to turn clay into the opposite of what it normally is in soil: positively puffy. Expanded clay pellets are light, chemically inert, and nearly pH-neutral. They have good wicking properties, hold roughly their own weight in water, and drain quickly. The pellets are sold under various brand names but are more difficult to find than most other substrates. They are also heavier and more expensive than other substrates.

Rock wool: Like sugar in cotton candy, rock wool's underlying minerals are melted at a high temperature and then spun out into fibers. Rock wool can be shaped into seed-starting cubes, blocks, slabs, or loose granules, and the water-retentive properties vary according to the particular mineral used. Rock wool is mostly used for seed starting (as a sterile medium with high water and air capacity) and in hydroponics. The fibers in it can be made to be hydrophilic (attracting water) or hydrophobic (repelling water), allowing blends of the two to produce a wide range of very specific water-holding capacities.

Nutritive

Chemical fertilizer is like instant gratification. Sure, it hits the spot. But will it still love you in the morning? Next season? Three years from now? Very often, it won't. There are a number of natural, nutritive abiotic soil amendments, however, to improve soil in need of fine-tuning over a longer term. Many need to be applied

A quartet of superheated abiotic amendments (clockwise from top): perlite, vermiculite, expanded clay, and rock wool.

Sumiballs may not be great for large plantings, but they're good for rooting cuttings, such as this Vietnamese coriander.

more heavily than chemical fertilizers because they release their nutritive value slowly over several years. In fact, some—such as greensand and granite dust—release their nutrients so slowly (if at all), that you should opt for the finest grade if you're going to apply them. If you want to use any of these substances in organic agriculture, you should first check with the Organic Materials Review Institute (OMRI) to make sure that they are permissible. Some have synthetically made doppelgangers, and some are permissible only if your soil has a documented deficiency.

A big recommendation of many organic farmers is to "remineralize" the soil. The thought is that we have been robbing soils of nutrients since the dawn of agriculture. As geologist Ward Chesworth of the University of Guelph remarked in the *New York Times*, farmers have been "making tea with the same bag for 10,000 years." This extraction has only accelerated since Baron Justus von Liebig identified the big three plant macronutrients (nitrogen, phosphorus, and potassium) in the mid-nineteenth century. With N-P-K, the new be-all-end-all of agriculture, organic matter and trace elements—once nurtured through traditional farming practices—were relegated to being the nutritional equivalent of yesterday's news. As farmers have rediscovered their importance, there's been a new emphasis on letting them out of the attic to restore them to the soil through an assortment of naturally occurring minerals.

Rock dusts: Also called *flours* or *meals*, these are one proposed solution to the demineralization of the soil. The effect depends, of course, on the particular rock used. Granite dust, for example, is about 3 to 5 percent potassium and two-thirds sand (silica) and has about nineteen micronutrients. Crushed basalt or basalt dust contains many minerals, as well, and may be more readily available as a by-product of the road construction business.

Azomite: This is not technically a rock dust; it is mineralized, compacted volcanic ash (66 percent silicon dioxide, 11 percent alumina). It is used in a similar way, however, containing about 2.5 percent potassium and more than seventy trace minerals and having a pH of 8.0.

Greensand: This is a sandstone believed to be the sedimentary remains of ancient shallow oceans. It is usually (but not always) green. The color comes from a mineral called glauconite, which is actually a sand-size clay. That doesn't seem to make sense to me, either, though this unusual state of affairs apparently gives greensand some desirable properties. It can hold much more water than regular sand, loosens clay soils, and aggregates sandy soils. It's about 5 to 7 percent potassium—which it releases very slowly—and is often used as a source of that macronutrient in organic fertilizers. It also contains calcium, magnesium, and about thirty micronutrients, and it improves a soil's ability to store these elements.

Langbeinite: Often sold as Sul-Po-Mag or K-Mag, this is a minimally processed mined ore that is 22 percent potassium, 22 percent sulfur, and 11 percent magnesium. Unlike many other naturally occurring nutritive abiotic amendments, it is soluble and quickly available, so it should be used sparingly.

Gypsum: You may recognize this as the chalky stuff in drywall, the chalky stuff in plaster of Paris, or, indeed, the chalky stuff in chalk. It's about 22 percent calcium and 17 percent sulfur and has a host of specialized uses: loosening clay soil, kicking up a soil's calcium and sulfur levels without affecting pH, and neutralizing excess sodium or magnesium. It's also called "land plaster."

Epsom salts: Epsom salts are about 10 percent magnesium and 13 percent sulfur. Like langbeinite, they are highly soluble—that's why they are used in bath salts, after all. They are also used in foliar sprays. If you're going to apply them to plants instead of baths, use them under advisement, not willy-nilly.

Rock phosphate and colloidal (or black rock) phosphate: Both provide phosphorus (in its available phosphate form, P_2O_5), calcium, and about a dozen trace elements. Although both mined products are very high in phosphate—about 30 percent and 20 percent, respectively—only about one-tenth of that amount is immediately available; in fertilizer code, they'd be about 0-3-0 and 0-2-0. For the less patient farmer, there is also superphosphate (0-20-0) and even triple superphosphate (0-45-0), which both involve phosphate rock that has been chemically treated with acid to make the phosphorus more available.

Lime: This is another amendment that takes confusingly many forms—burnt and slaked, powdered and pelletized, calcitic and dolomitic. Whatever its source, lime has one main purpose: increasing pH (making the soil less acidic). It may also have a secondary purpose (adding calcium) or a tertiary one (adding magnesium). Three varieties should suffice for most urban-farm needs. Oyster shell lime, or aragonite, is 96 percent calcium carbonate and derived from oyster shells. Calcitic and dolomitic lime both contain calcium and magnesium; dolomitic has slightly less calcium and significantly more magnesium. Generally speaking, dolomitic limestone suits sandy, well-drained soils (which don't maintain calcium and magnesium well), and aragonite is good where there's a magnesium surplus. Calcitic lime is probably best somewhere in between. Ask your local cooperative extension. Sometimes dolomitic lime is better for clay soil, and sometimes calcitic is better. Pelletized lime is easier to apply (and more slowly released) outside, though you might want the finer-grain (and cheaper) kind if you're just using it in containers or small plantings.

Biotic

Biotic matter includes all living or once-living things. Carbon is a common denominator in all living things, from acorns to viruses to us. We all decompose, as well, with the living stripping down the dead. We're a resourceful bunch, living things. Many of the best available soil amendments are biotic, although not all are equally sustainable from an environmental standpoint.

Sphagnum Moss and Peat Moss

Both sphagnum moss and sphagnum peat moss come from the same source: bog-loving mosses of the genus *Sphagnum*. Sphagnum moss is harvested from living

A one-liter brick of compressed coir (left) produces about 8 liters of loose coir when rehydrated (right).

plants, while sphagnum peat moss (often just "peat moss") consists of compacted and partially decomposed sphagnum moss. Both versions are favored because of their slow decomposition rate and tremendous capacity to hold water; peat moss (but not sphagnum moss) is also virtually sterile. Peat moss is very acidic (pH 3.5 to 4.0) and would require the use of lime or other amendments in most circumstances.

Although living sphagnum moss can be harvested sustainably—as is done most notably in New Zealand—peat moss can take hundreds of years to become harvestable. Peat bogs also harbor wildlife and, according to one study, sequester over 450 billion tons of carbon—equal to decades' worth of Earth's industrial emissions. Preserving these carbon sinks makes sense. My own inclination is to favor

A Question of Safety

Those of you who flee from biotic soil preparations for the sterile safety of completely soilless mixes might be surprised to learn that these mixes can prove more dangerous than microbe-teeming soil—at least to the farmer, if not to whatever they plan on growing. Dry perlite is a "nuisance dust" that can irritate eyes and affect breathing, so goggles and a face mask should be used when handling it.

Vermiculite is another nuisance dust, and some vermiculite from the once-largest mine in the United States (now closed) was contaminated with asbestos. Though it is now obtained from other mines, and apparently tested regularly, taking the same precautions as you would with perlite makes sense. Ditto for rock wool.

When mixing or otherwise handling these amendments, also consider misting them with a spray bottle to keep down dust. And since you're buying gloves anyway, using them while handling sphagnum moss is a good idea because it can host a rare fungal contaminant.

sphagnum moss over sphagnum peat moss if you can, and to favor coir (see opposite page) over both.

Note that although peat moss and sphagnum peat moss can hold a tremendous amount of water, they can also dry out easily. If that happens, they go from hydrophilic (water-loving) to hydrophobic (water repelling), appearing to punish you by becoming seemingly impossible to remoisten. The solution is to soak the moss in water overnight, preferably with a drop or two of vegetable-based dishwashing detergent (or castile soap) to act as a wetting agent; you can then squeeze out the excess water. Coir suffers the same Jekyll/Hyde characteristic, but to a lesser extent.

Coir and Coir Pith

Coir is the fiber in coconuts, and coir pith (or coir dust) is a by-product of coir-fiber production and much more sustainable than sphagnum moss. It also has a pH closer to neutral (5.5 to 6.8), lasts longer, and has better water-retention capacity.

The downsides are that it is more expensive, is often treated with salt water in production, and may have relatively high levels of potassium, sodium, and chloride.

It's not a bad idea to soak and squeeze coir an extra time or two in case there's residual salt in it, and to tweak fertilizers for coir-based mixes to contain more nitrogen and less potassium. Some newer producers of horticultural coir take extra efforts to eliminate salts. Coir is often compressed into seeding disks or pellets or into bricks that expand dramatically when you add water.

Wood Bits

Composted pine bark is often used to lighten soil mixes in the eastern and midwestern United States, and fir bark is used in the western states. Either can be used in place of peat moss in some blends, although they are often combined; this is probably more ecologically friendly since bark is a by-product of the lumber industry. Both pine and fir have a pH of about 5.0 to 6.5, and—like other carbonaceous items—will likely drain some local nitrogen, so nitrogen supplementation should be a consideration (as should liming) if your crops require fairly neutral substrate.

Composted sawdust would seem very environmentally friendly, and it certainly can be, but not all sawdust is created equal. That from incense cedar, walnut, redwood, western red cedar, and peat-grown conifers, for example, can be toxic to plants. Would you be able to distinguish one chip from another? I'm not saying don't use it; just know your supplier and test it out on a small scale first.

Leaf Mold

Leaves are one resource that literally falls from the sky in many places, providing a free starting point for leaf mold, which comprises partially decomposed leaves. Leaf

Urban Homesteading

Kelly Coyne and Erik Knutzen of Los Angeles know a thing or two about soil. Heck, they list "Build your soil" as tip number three in "The Seven Guiding Principles of Successful Urban Farming" from their bestselling book The Urban Homestead. Without enough kitchen scraps to make a big dent in their soil, which they manage organically, they have kept an eye out for vegetable waste at farmer's markets, composted the droppings of their own small flock of chickens, and scouted out sources for that carbon/nitrogen-ratio superstar— horse manure—produced in droves at the City of Angels' many equestrian centers.

About twelve years ago, the pair bought a house and began the shift from being dependent consumers to self-sufficient producers. They grow crops, keep bees, husband chickens, recycle water, pickle and preserve, forage the neighborhood, cultivate neighbors, and even brew their own hooch—kind of like Ma and Pa Ingalls, but within two blocks of Sunset Boulevard. And instead of a pig-tailed "Half-Pint," they have a Doberman Pinscher named Spike.

Erik and Kelly also blog religiously (check out www.rootsimple.com), which led to the book deal. Now they are pretty much full-time urban farming gurus, sharing their experience through the blog, books, demonstrations, and other means. Starting your own urban farm will probably not lead to a book deal, but that doesn't mean you can't learn from their experience—hence the title referencing their seven principles or, as they put it, "what we've learned the hard way."

Here's what they've learned:

1. Grow only useful things.
2. Region matters. A lot. (Erik notes, for example, that they live in a "Mediterranean" climate, shared by only a tiny fraction of the land on Earth. So they simply disregard anything on the back of seed bags, since it's probably not appropriate for them.)
3. Build your soil.
4. Water deeply and less frequently.
5. Work makes work. This is a permacultural ethic advocating one to work with nature because the alternative promises you a long, hard, and ultimately regretful slog.
6. Failure is part of the game.
7. Pay attention and keep notes.

mold provides fine biotic matter and trace elements. Leaves have a very high carbon content and therefore take considerably more time to break down on their own than does nitrogenous material such as grass clippings. Oak leaves and pine needles are somewhat acidic and will acidify the soil— making them good dressings for rhododendrons and other acid-loving plants— but fully composted leaves result in a near-neutral pH level. Many places now compost leaves at the municipal level and make the resulting leaf mold available for free or for a nominal fee, so it's worth checking with your local public works office. Just note that for all its potential for aerating and retaining moisture in soil, leaf mold should not be used as a substrate in itself—it can contract and harden into something like cement. Always mix it in with existing soil or a heaping amount of abiotic ingredients, such as sand.

Leaves and leaf mold also make excellent additions to compost because the large surface of leaves encourages microbial activity. This might be the ideal use of leaves, in fact, because the hot phase of composting (see "Composting How-To" later in this chapter) should destroy any pests harbored by the fallen leaves.

Manure

One of the most tempting aspects of using animal manure these days is that there is just so much of it—if it were divvied up annually, we'd each take away more than a ton. If all of the nitrogen and phosphorus in this waste were removed instead, these elements could provide a significant fraction of the fertilizer consumed each year. No wonder pre-Industrial farmers paired plants and livestock so effectively, and why manure has often been a valuable commodity—even from us.

"Night soil," a euphemistic term for human manure, was once commonly used for fertilizer. It has since been largely phased out because of improvements in sewage infrastructure, concerns about disease avoidance, better employment opportunities (who aspires to be a night-soil recycler?), and the notion that it's incompatible with modern civilization. Interestingly, though, the Food and Agriculture Organization (FAO) of the United Nations and other organizations have proposed reintroducing it in some contexts—not indiscriminately, but where nutrients are scarce (or expensive) and the night soil is treated (hot composted) to eliminate or minimize potential pathogens and parasites. (Its N-P-K nutritional profile, incidentally, is

1.2- 0.35-0.21.) New York City generously contributes about 1,200 tons of biosolids daily, for example, which are sent to do everything from fertilize median strips to restore Texas prairieland to fertilize citrus fields in Florida. Think about that over your morning orange juice.

Green Manure

Sometimes plants are grown specifically for the purpose of being sacrificed for the soil, and that's what green manures are. The plants may be chosen for the biomass that they will add to the soil, for deep roots that will break up soil and bring up nutrients, or for the ability to fix nitrogen from the air into the soil. Alfalfa serves all three purposes, but farmers dabble in a huge range of green manures, both leguminous (including alfalfa, clover, and vetch) and nonleguminous (including millet, rapeseed, and sorghum). Combinations, such as peas and oats, are often used. At some stage of growth—usually before flowering— the plants are turned into the soil.

One of the wonderful aspects of these crops is that they can serve at least two functions. The first, obviously, is to enrich the subsoil by becoming green manure. The second is to enrich the environment above the soil by suppressing weeds, "trapping" bad and attracting good insects, and protecting the surface soil from nutrient leaching, erosion, and the elements. If the latter is the primary purpose, these plants are usually called cover crops. Whatever you call them, green manures or cover crops, plants such as alfalfa have the potential to achieve both ends for you.

Meal of Blood, Bone, Feather, and Fish

Sounding very Churchillian, this quartet of animal by-products comes from cows, pigs, chickens, fish, and assorted other unfortunates. Sold separately or combined in organic fertilizers, each has its agricultural niche. Blood meal and feather meal provide nitrogen (both about 12-0-0, with rapid availability in blood meal and slower availability in feather meal), bonemeal provides primarily phosphorus (about 2-14-0) with a big helping of calcium, and fish meal (about 9-5-3) provides decent nitrogen, phosphorous, and trace nutrients.

Blood meal and fish meal are both acidic; bonemeal is alkaline. These materials are often available separately (at least in the case of blood meal and bonemeal), and organic fertilizers usually include one or more of them. Blood meal and bonemeal are impermissible under biodynamic standards, however.

Compost

The aforementioned materials are just opening acts for the main event in biotic amendments: compost. All of them—not to mention newspapers, coffee grounds, and most other once-living things—can be composted successfully. Composting is essentially a deliberate attempt to make humus, and compost is the holy grail of substrates for many farmers—the closest that gardeners get to pixie dust. It is, in fact, kind of magical. No matter what the pH of materials being composted, the finished product is always just about pH-neutral—and even buffers very acidic or alkaline soils on which it is placed. Bad-smelling ingredients become sweet-smelling soil. Good compost can actually suppress plant diseases and even render toxic metals in the soil harmless by chelating them (though I wouldn't rely on this alone if you know that your soil is contaminated).

Compost can be purchased readily enough, but making your own can be easier, cheaper, and more ecologically sound. Although DIY composting may raise the specter of "effort" for people who have heard rumors of carbon/nitrogen ratios, pH balances, and the differences between aerobic and anaerobic decomposition, it is well within the ability of anybody with "a skill for sandwich making and the possession of a few guiding principles," according to Dr. Judyth McLeod, an Australian expert in sustainable horticulture. Look for "OMRI Listed" on compost—signifying approval by the Organic Materials Review Institute—if you want to be sure that it is permissible in organic farming.

Composting How-To

Things fall apart. So will you one day. I'm barely here now. Don't worry that you won't know every possible thing about composting. It's a forgiving science. Whole books could be devoted to the subject—and, indeed, whole books are.

While there are many different composting methods, from Bokashi to Hügelkultur to vermicomposting, the fact is that your pile of biotic materials will eventually become compost even if poorly made and poorly tended. The design of your composting vehicle is similarly accommodating. Although there are a wide

Urban Composting

Composting really embodies urban-farming ideals in that it recycles a city's own waste into a means of feeding itself, closing the loop. It cuts down on transporting fertilizer into the city and solid waste out of the city—both likely by greenhouse gas-belching trucks. It also reuses solid waste productively instead of letting it rot in a landfill. Aerobic decomposition in composting produces mainly carbon dioxide, while anaerobic decomposition in landfills produces mainly methane, which is much worse for the atmosphere. The scale of the savings is not to be underestimated: the average world urbanite produces a bit less than 1.5 pounds (0.7 kg) of trash per day, about half of which is biodegradable. According to the EPA, the average American produces about 4.4 pounds (2 kg) of trash per day, about three times the average of city dwellers around the world.

Applied to Peoria, Illinois (pop. ~113,000), the American average would translate yearly to about 90,740 tons of trash. If converted to a liquid for the sake of comparison, that would be roughly equivalent to all of the milk that the United States produces in a year, all of the ethanol that the world produced in 2009, or enough water to make an ice cube 142 feet (about 43 m) on all sides. Unlike milk, ethanol, or water, however, trash is not eminently useful and may, in fact, be very detrimental if not properly handled. And that's just for Peoria, which has less than .037% of the total population of the United States.

Composting is often done in side-by-side bins: one for fresh waste, the other for finished compost to use.

variety of established techniques, such as the Indore Method, the Windrow System (and its California variation), the New Zealand Box, Ogden's Step-by-Step Method, and the German Mound—all bringing to mind exotic chess openings. The fact is that things rot. Just try to stop them.

Three basic methods are easily employed by urban farmers: vermicomposting, bin composting, and sheet composting.

Vermicomposting

Vermicomposting involves yoking for services that most lowly of livestock: the worm—specifically red worms, *Eisenia foetida*, *E. andrei*, and *Lumbricus rubellis*. These worms are not earthworms, and, for the most part, they cannot even survive outside in the ground. But they are composting superstars. They can eat 50 to 100 percent of their weight in biotic trash each day. So 1,000 worms—about 2 pounds (0.9 kg)—could devour the roughly half-pound (quarter kilogram) of food scraps

Two redworms devour an old peat pot and kitchen scraps. Those small yellow orbs are cocoons—more worms to come!

that the average person generates each day, as well as a chunk of the daily pound and a half of paper, too. They multiply quickly, so if you have more worm slop to give them, you'll soon get more worms to eat it. Avoid giving them meat or dairy products; anything really fatty, oily, or salty; pet waste; or much onion or garlic. Apparently none of these is healthy for the worms to eat, and some also pose the risk of attracting vermin.

The worms prefer a temperature of about 60 to 75 degrees Fahrenheit (15.5 to 24 degrees Celsius)—probably about the same as you—but can take up to 20 degrees Fahrenheit (less than 1 degree Celsius) in either direction. Managed well, a vermicomposter will produce little if any discernible odor. There are many commercial worm bins available, most of which resemble little pagodas made of shallow nesting trays, often with a spigot to tap the liquid runoff ("compost tea," which can also be produced by soaking finished compost in water) resulting from vermicomposting. Several commercial varieties are readily available, but you can cheaply make one yourself.

The easiest way to make your own worm bin is with an opaque plastic bin in the 6- to 12- gallon (23- to 45-liter) range. Long, wide, and shallow bins are preferable to deep ones. Perforate the bottom with a bunch of holes, using either a nail or a ¼-inch (64-mm) drill bit. Perforate around the top rim of the bin with slightly smaller holes. The bottom holes are to let liquid (compost tea) out and some air in, and the top holes are all for air; the top holes are smaller to deter flies and other pests. Order some worms online, or scope out a local bait shop that sells them. In the meantime, fill the bottom of the bin with a few inches of wet, crumpled newspaper, moistened coir, or sphagnum moss. This will be the "bed" for the worms. At this point, the bin is worm-ready.

When you get the worms, place them on the bed and let them acclimate for a day or two. You can also leave a small amount of food waste in there with them. After a few days, you can start giving them small meals, each time burying the slop in a different part of the bin. See how it goes, and ramp up your feedings slowly. Very slowly. I, for one, expected my worms to act like some kind of living garbage-disposal unit right after their journey through the postal system. I ended up overfeeding them, which produced little vermicompost but many fruit flies. Go slow...and it wouldn't hurt to keep some flypaper around at first.

The worms' bin should always be covered to keep it dark and moist for the worms. You should also have some kind of tray underneath the bin to catch the compost/worm tea as it slowly dribbles out. The neatest solution, described very clearly by Washington State University's Whatcom County Extension, is to buy two bins with lids. Perforate the bins as previously mentioned and then add extra-small holes to one of the lids. Then you can use the unpunctured lid as a catch tray for worm tea, the perforated lid as a cover, and the second bin as a tag-team partner.

In other words, when the first bin is almost full, place the second bin right on top of the compost (without the lid), and put some new bedding and food in the second bin. The worms will eventually migrate through the holes into the second bin. Once enough have, usually after a month or two, make the second bin the bottom bin. The first bin will be filled with vermicompost to use. When you've emptied out all the compost and the second (now bottom) bin is almost full, place the first bin on top of the second with fresh bedding. And so it goes, from the first bin to the second to the first to the second, ad infinitum. Commercial bins work in the same basic way.

Bin Composting

The second big method for urban farmers is bin composting, which is a step more sophisticated than just throwing everything into a heap. It's a contained heap. The bin or container can be constructed out of just about anything: wood, unmortared cinder blocks, wire mesh, commercial plastic bins, extra fencing, and so on. You have three guiding principles, however.

1. Whatever it is, it should let in plenty of air.
2. The bigger the better. A cubic yard (27 cubic feet) is generally considered the minimum needed to achieve hot composting, which achieves temperatures high enough to kill weed seeds and many plant pathogens. You can certainly have smaller-volume bins, but the compost will not get hot enough to eliminate all pests, nor will it finish as quickly. It will still work, however.
3. Aim to keep out vermin (such as rats). Wire-mesh bins should employ small-holed hardware cloth, for example, rather than chicken wire. Hardware cloth also makes sense for any ground-based bin's bottom (it's too finely knit for mice, but not for worms) as well as to surround wood or plastic bins. A galvanized steel trash can with some air holes punched into it works well for cool composting, but it lacks the volume needed for hot composting.

Variations on a compost bin (from top): a steel mesh cylinder, a galvanized trash can, and a plastic commercial composter. All three will get the job done, but only the commercial one is likely to create a "hot" pile.

Irrespective of what type of bin you choose, the rumors about balancing C/N ratios are true. All biotic things have a C/N ratio, which expresses the percentage of carbon (always higher) relative to the percentage of nitrogen. Coffee grounds, for example, are about 20:1. Freshly fallen leaves are in the range of 40:1, depending upon the type of leaves and other factors. A 30:1 C/N ratio is generally considered the ideal target for composting. Materials above that are considered carbonaceous, and materials below that are nitrogenous.

The task of the composter is often described as balancing the "browns" (carbonaceous materials, which are often, but not always, brown) with the "greens" (nitrogenous materials, which are often, but not always, green) to achieve the golden ratio. Don't think of it as chemistry (carbon and nitrogen). Think of it as a Stendhal novel or Kurosawa film, with hordes of browns and greens meeting in a breathtaking skirmish of mutual destruction. Imagine yourself as a field marshal or perhaps a Broadway choreographer. Or maybe you should just stick with the chemistry.

A simple "cold" pile is not too picky about the C/N ratio, but it will decompose more quickly the closer it is to 30:1. Smaller ingredients also accelerate the process because they provide a greater surface area for microbial action. Turning, pitchforking, or otherwise aerating the pile will help, too. In fact, if you have a near 30:1 C/N ratio, all of your ingredients at hand and in small pieces (shredded leaves, wood chips, and so on), a commitment to turn the pile regularly, and about a cubic yard of material, you'll probably soon have a hot pile. And a hot (or "managed") pile can turn garden leftovers into compost within as little as two weeks.

Concocting the Perfect Compost

But enough science. Let's talk about getting our hands dirty. Think you can find a single, pre-packaged ingredient with a perfect 30:1 ratio? Horse manure! That's right, horse manure has just about a perfect 30:1 ratio, and maybe—just maybe—you can find that somewhere in your city (stables for mounted police?).

Other materials that come very close are a cranberry-filter cake (31:1), olive husks (30–35:1), potato-processing sludge (28:1), and fish-breading crumbs (28:1). So, let's face it: most of us will have to juggle rather humdrum ingredients, offsetting carbonaceous stuff such as an old phone book (772:1) with nitrogenous ones such as the day's coffee grounds (20:1).

It's good to mix up materials to provide a range of trace elements. Remember that the 30:1 ratio is a rule of thumb rather than a mandate. Thankfully, you don't need a lab to tell if your mix of carbon and nitrogen is working out; you can just use your senses.

- Smells bad and is slimy or wet: add dry, carbonaceous materials and turn for aeration.
- Smells bad but is not slimy or wet: add coarser materials and turn.

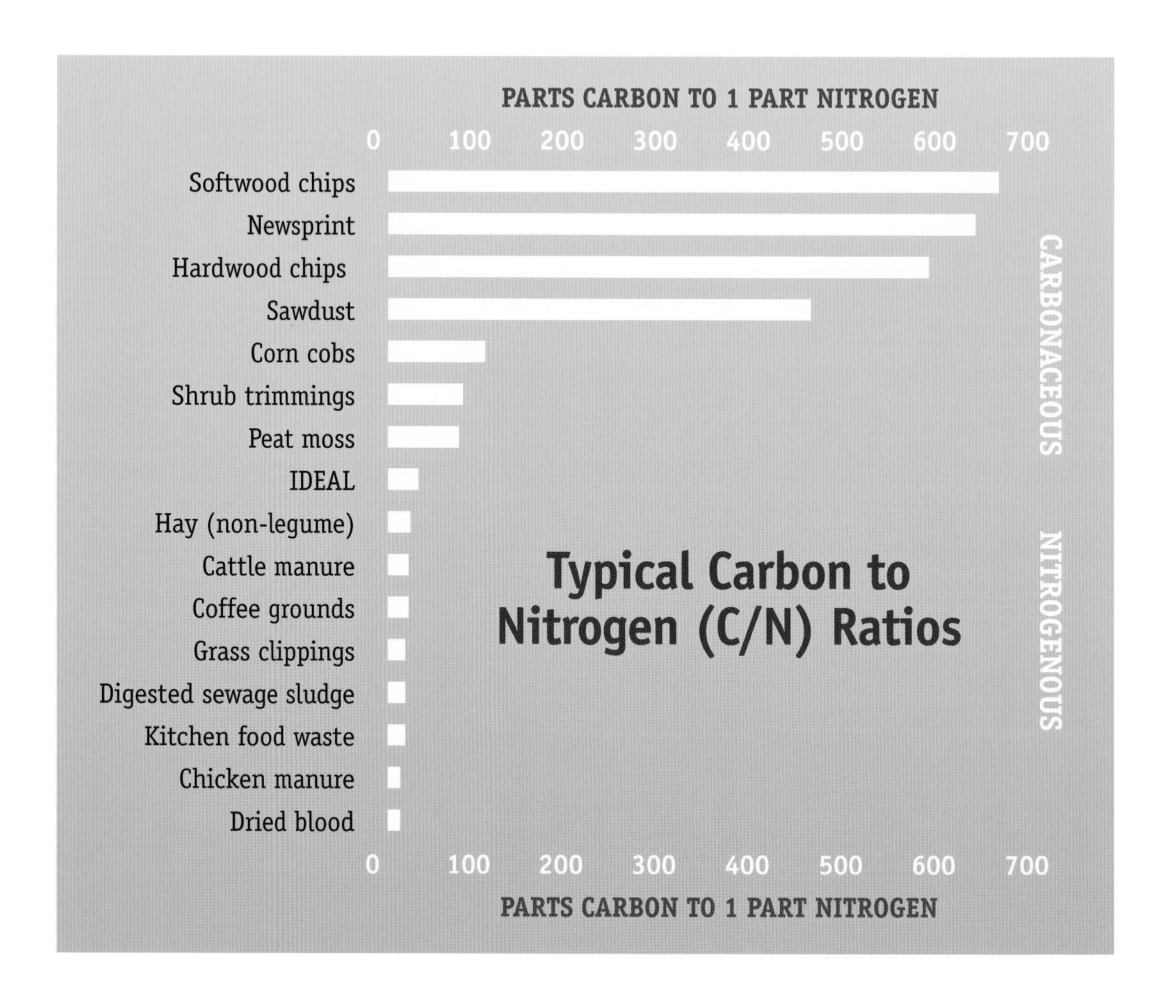

- Smells good (earthy) but is not heating up and is a small pile: make the pile bigger—8 cubic feet at the bare minimum (2 feet by 2 feet by 2 feet [61 cm by 61 cm by 61 cm]) but preferably at least a cubic yard (3 feet by 3 feet by 3 feet [91 cm by 91 cm by 91 cm]). Or do nothing— you'll just need to wait longer.
- Smells good (earthy) but is not heating up and is a big pile: add nitrogenous materials, turn, and dampen if somewhat dry.

Moisture is an important consideration, especially in a hot pile. The decomposing microorganisms love moisture, but the exposure to air and the heat from the center of the pile risk drying out materials on the periphery. In that case, spray the pile with a hose or otherwise water it. If your pile gets too wet for some reason—which might make it smell, too—there are several solutions. You can add dry material (such as leaves) and turn the pile.

If rain is in the forecast, cover the pile until it reaches a more

Hot Composting

Like a concerto of decay, a well-made, well-attended hot pile goes through three acts. The pile heats up to between roughly 100 and 120 degrees Fahrenheit (38 to 49 degrees Celsius) during the first act, whose featured soloists are mesophilic bacteria, which thrive in some warmth. Those ingredients that are easily broken down do so during this stage.

Heat-loving thermophilic microorganisms are the stars of the second act, when temperatures climb beyond 120 degrees Fahrenheit (49 degrees Celsius), peaking at between about 150 and 170 degrees Fahrenheit (about 66 to 77 degrees Celsius). This is hot enough to burn your hand (if you decided to plunge it to the center for some reason), break down cellulose, and kill seeds and many disease pathogens. Countless good organisms are killed, as well, but their kind will make a triumphant return in the third act.

The third act is called "curing" and is characterized by cooling off, increased humus production, and the recolonization of organisms that couldn't stand the heat.

There are a variety of washable pails for kitchen scraps available with charcoal filters to eliminate smell.

desirable moisture level. Finally, you can insert perforated PVC pipes or similar tubes into the center of the pile; the extra exposure to air should help it dry more quickly.

Sheet Composting

If you have some farmstead on the ground, such as raised beds, sheet composting is another way to go. Just like it sounds, it involves laying down materials to be composted in a sheet. For example, you might lay down a few sheets of newspaper, then cover the newspaper with nitrogenous materials such as grass clippings and kitchen waste (coffee grounds, vegetable peels, and so on), and then cover that with fallen leaves. Wet it all down. You can add to the sheeted area as materials become available, either increasing square footage or just layering again above the existing sheet (hence the common description *lasagna gardening*).

Some farmers till under the layers once they have partially decomposed to further mix the ingredients and speed the incorporation of compost into the soil. Another option is to plant a green manure, till it under, and then cover it with carbonaceous materials and whatever other layers you want to add. Neither action is necessary. As is, sheet composting is a slow builder of soil, adding organic matter and nutrients very gradually. If you are lucky enough to have sufficient farmstead to leave some of it fallow, sheet composting is a great way to improve the soil year after year. If you have five raised beds, for example, you can plant in four and use the fifth for sheet composting. Next season, you can plant in your composting bed and use a different one for sheet composting. It's a convenient way to recycle biotic materials that are enough for a sheet if not a heap and to restore the quality of the soil throughout your farmstead. Sheet composting can also suppress weeds, if done carefully, and bring many of the same benefits as the next topic: mulch.

Mulch

Mulch is a multipurpose ground cover. It holds moisture in the ground while protecting the topsoil from rain and wind erosion. By insulating the ground, mulch moderates temperature extremes, which is particularly important in cold areas where freeze/thaw cycles can heave plants out of the earth. Mulch suppresses weeds and mitigates ground compaction. If composed of biotic materials, as sheet compost is, it even helps to nourish the soil. Yet it can also be made of abiotic material, such as plastic, shredded rubber, or pebbles. These don't improve the soil and may not provide the same level of insulation (depending upon the material), but they still offer the other protective advantages.

People in rural areas often have access to straw and hay for mulching. Chances are, you won't. Bark chips for mulch are readily available commercially, and wood chips or leaf mold might be available for free at a municipal recycling center. You might have other fluffy, biotic, local choices in your state—such as rice hulls in Arkansas, cocoa shells in Wisconsin, or pine needles in Florida—so look around.

Nitrogen

One issue that affects the use of biotic-sheet composts and mulches is nitrogen depletion. Because they tend to be so high in carbon, materials such as wood chips, newspapers, or leaves take in local nitrogen as they decompose. Adding a high-nitrogen amendment such as blood meal can help keep nitrogen available for crops.

Spring and fall are generally the best times to apply mulches. Gardeners in cold areas should avoid applying mulch just as it starts getting very cold (so rodents don't claim it as home) and when it is already very cold (because the insulation will delay the warming of soil in springtime). You should also leave a few inches of unmulched perimeter around every plant to avoid giving cover to slugs or encouraging fungal attacks.

Pebbles or pea gravel make good mulch for plants in containers. Another option is a low-growing innocuous plant that is not very thirsty, such as sedum.

Soil Mixes

The easiest solution, especially if you're planting small-scale, is to buy a commercial soil mix. Soil mixes are readily available, not too expensive, and well calibrated in their balance of water and air, nutrients and pH, and biotic and abiotic matter. Potting soil or potting mix is usually the way to go with container plantings. It tends to be light, well-drained, nutrient-rich, and composed of things such as sphagnum moss, perlite or vermiculite, and composted animal manure—frequently a "soilless" mix. Garden (or planting) soil (or mix) tends to be somewhat heavier but still well draining, high in biotic matter, and filled with nutrients. It is often made specific to certain kinds of plants (vegetables, citrus, ornamentals, and so on), and is meant to be mixed with existing topsoil. It's probably the way to go with most raised beds. Commercial topsoil is generally the heaviest, cheapest, and least porous. Still, it may be better than your own topsoil. It is intended mainly for filling in depressions and improving your own soil.

Another option is to buy commercial soil products and blend them yourself. Try to do this outside, away from wind, and with the ingredients lightly moistened first. Keeping a spray bottle at hand to dampen any dust is probably a good idea, as is wearing a mask and gloves. Most mixes contain something to retain moisture (peat

Recipes for Blending Your Own Soil Mixes

Basic Potting Mix

1 part topsoil (your own or store-bought)
1 part biotic matter (compost, leaf mold, coir, or peat moss)
1 part sand, perlite, or vermiculite (or combination thereof)

Sterile Seed-Starting Mix #1 ("Peat-lite")*

1 part vermiculite
1 part sphagnum peat moss
*These ingredients, combined with dolomitic lime and fertilizers, comprise the classic "Cornell Peat-Lite Mix A."

Sterile Seed-Starting Mix #2

1 part vermiculite
1 part perlite

Microbial Seed-Starting Mix

2 parts vermiculite or perlite (or combination thereof)
1 part fine compost or leaf mold

Cuttings Mix

1 part vermiculite or perlite (or combination thereof)
1 part sand
1 part sterile soil or peat moss (for sterile version) or fine compost (for microbial version)

moss, sphagnum moss, coir, or other biotic matter), something to increase porosity (perlite, vermiculite, or sand), and something nutritious (compost, leaf mold, or biotic/abiotic fertilizers). Lime is often added to naturally acidic peat mixes to balance the pH. Note that the compost should always be mature, the vermiculite almost always "horticultural," and the sand almost always coarse (such as builder's sand—not fine or tropical sand). You can use finer grades of vermiculite and sand in seed-starting mixes.

Soilless Substrates

There may be as many fans of soilless media as there are of soil. Some of the soilless options are organic (in the chemical sense, meaning that they contain carbon and used to be alive) and some are inorganic (may or may not contain carbon and are abiotic). Many packaged and prepared soilless media contain a blend. In fact, all three seed-starting mixes previously mentioned can be considered "soilless." Two out of the three are sterile, and sterility is a common point in favor of soilless media, particularly when plants are at their most vulnerable. Yet soilless media are sometimes favored for the whole life cycle of the plant. At the United States' Antarctic research facility at McMurdo Station, for example, perlite/vermiculite blends are used to grow vegetables and herbs, in part to avoid contaminating the pristine Antarctic environment with foreign soil microbes.

McMurdo Station grows its food in the simplest kind of hydroponic set-up: the static solution method. Plugs holding the plants are immersed in a bucket of nutrient solution. The perlite and vermiculite give the plant roots something to grab onto and facilitate aeration. In fact, all of the nonnutritive abiotic soil amendments previously discussed are used as hydroponic substrates, as are some of the biotic ones (coir, for example). Some methods of hydroponics, such as the nutrient film technique (NFT), don't even use substrate as we'd commonly think of it. In NFT arrangements, slightly angled channels made out of PVC pipe or other materials allow a constant shallow trickle of recirculating solution to reach the bare roots of plants, which are inserted into holes on the top of the channels (pictured opposite).

Aeroponics takes hydroponics a step further, maximizing root exposure to air as well as water and nutrients. It employs a high-pressure mist of nutrient solution to

One of the advantages of hydroponics is never having to wash soil off of your produce.

feed a plant's bare roots. Aeroponics uses even less water than hydroponics, which is itself very water-efficient. NASA has experimented with aeroponics for space travel, and there are many aeroponic kits commercially available, from a $100 unit to grow herbs to entire turnkey aeroponic farms.

Green roofs also usually employ soilless substrates. It's a tricky situation. The substrate can't hold too much water because of the load stress on the roof, yet it has to hold enough for plants while simultaneously letting air get to the roots. It also has to be light itself to avoid putting too much pressure on the roof, but not so light that it blows away. The green roof on Chicago's City Hall, for example, weighs about half per cubic foot what topsoil does.

Germany has been at the cutting edge of green-roof design for decades, and best practice calls for following the design guidelines of a German nonprofit, Forschungsgesellschaft Landschaftsentwicklung Landschaftsbau (The Landscape Research, Development, and Construction Society), known, for obvious reasons, as FLL. Extensive or intensive, single layer or multilayer, soil or bulk material, a green roof's substrate is subject to exacting specifications by FLL. There are many proprietary substrate mixes for green roofs, often calibrated to the type of roof and location. They usually combine a lightweight aggregate material with organic matter.

BTM Optigreen extensive one-layered substrate type M, for example, contains lava rock, pumice, expanded slate and clay, sinter slag, crushed brick, manure, and composted green waste.

Like Building an Onion

The planting substrate is just one factor in a green-roof building practice that usually involves seven. A waterproofing membrane—modified bitumen or PVC would be typical—is laid down to protect the roof from leaks. A root barrier protects the waterproofing membrane from prying plant roots and may contain copper as a chemical root pruner. An insulating layer is often included just above or below the waterproofing membrane, establishing a basic level of insulation in winter, when the green roof provides little, and boosting the insulating effect in summer. A drainage layer aerates the roots and prevents water from puddling. It is sometimes combined with a retention layer to store some water. A geotextile filter fabric drapes over the lower levels to support the substrate. Substrate is added, and then, finally, plants are introduced. (It is not an undertaking for the weekend amateur. I'm fine with containers.)

Urban farmers are much more likely to grow plants in individual containers than on a whole field transplanted to the top of a building, but container mixes can be inspired by green-roof substrate if you need something particularly well draining (but moist) and light (but not too light). You might try, for example, polystyrene peanuts in the bottom fifth of the container, topped by a blend of 50 percent commercial potting mix, 30 percent expanded clay aggregate, and 20 percent mature compost. Or, if you can't obtain expanded clay aggregate, use a polystyrene peanut layer topped with a blend of 50 percent commercial potting mix, 10 percent coarse sand, 20 percent vermiculite, and 20 percent mature compost. Just make sure that you have enough drainage holes and that they are of such a size that they cannot be plugged by a packing peanut.

*****

So, soil or substrate? The way you answer reveals a lot about your horticultural philosophy. Either way you go, however—soil-based or soilless—substrate is what allows you to cultivate the main event: plants.

CHAPTER 6

Vegetable Matters

If you want to be a farmer, it pays to know a little something about plants. A lot, even. Plants are the main event for most farmers, and the more you know about them, the better you'll be able to profit from your efforts.

The taxonomy of plants can be sliced and diced many different ways, and one of these ways is based on life cycle. Annuals are plants that live a single year, produce seeds, and die. Most of the classic vegetables you'd find in a supermarket fall into this category, including broccoli, spinach, beans, peas, summer and winter squash, and lettuce. Biennials are plants with a two-year lifespan that often exhibit significant differences in their first and second years. The carrot, for instance, produces feathery leaves and a delicious swollen taproot in the first year. In the second year, it flowers, produces seeds, and dies. By the second year, the taproot we've grown to love has become woody and, if not exactly inedible, not quite as delectable. For this reason, many biennial plants are grown as annuals. In other words, they are harvested at the end of the first season, never allowed to reach the stage of flowering and setting seed. Obviously, some plants are allowed to live into the second year specifically to flower and produce seeds—otherwise, we'd have no more carrots!

PePPERS

Perennials are plants that grow for three or more years. Most, if not all, tree fruits are perennials; these include apples, cherries, pears, apricots, peaches, plums, and all citrus. So are most nuts, including almonds, pecans, walnuts, filberts (hazelnuts), Brazil nuts...basically everything you'd find in a can of mixed nuts, except for peanuts, which are annuals and grow underground. Many "berries" (such as blueberries, raspberries, and blackberries) are perennials, as well. Strawberries can go either way: some varieties are annuals, some are perennials. Perennial vegetables include artichokes, sunchokes, asparagus, and rhubarb.

A *hardy* perennial is one that will survive the cold season unassisted in a given climate, and a *tender* perennial is one that will not. A *half-hardy* perennial can survive the cold season with a little help. Fig trees, for example, are sometimes wrapped in burlap or laid down and partly buried in the northern United States.

For the most part, perennials will grow larger and more productive where they are hardy rather than where they are half-hardy. Trees and most shrubs are *woody* perennials, while their flimsier, pith-lacking cousins, such as alfalfa and most ferns and grasses, are *herbaceous* perennials. Herbaceous perennials characteristically die to the ground in winter and reemerge in spring, while woody perennials remain alive both above and below the ground. Even within the broad categories of herbaceous and woody, there are many ways to classify food crops, from phylogenetics (placing both apple and pear species within the subfamily Pyrinae, for example), to morphology (pomes, drupes, berries), to use (leafy greens, root vegetables), to seed form (monocot or dicot), to cultural requirements (light, soil type, pH, climate).

A Rose Is a Rose Is a *Rosa Sp.*

In elementary school, you probably memorized "King Philip Came Over From Great Spain" or one of its dirtier equivalents to remember basic taxonomic rank: kingdom, phylum, class, order, family, genus, and species. Plant taxonomy has only gotten more complex since then, with tribes, series, subphyla, and other ranks. Thankfully, cultivated plants are usually identified by just two or three names, the genus (its "generic" name), the species (its "specific" name), and the cultivated variety (known as the cultivar). That may seem much more complex than a common name, such as "rose," but it is actually very helpful. For one thing, there are scores of rose species with thousands of cultivars. For another thing, there can be tremendous variation within species. As any botanist loves to mention, for example, broccoli, Brussels sprouts, cauliflower, and kale are all the same species (*Brassica oleracea*).

In other words, there's almost no chance buying just any old rose will give you the fragrant French rose resembling peppermint candy you saw at the botanical garden, *Rosa gallica* 'Versicolor.' These scientific names follow a certain order: genus (*Rosa*), species (*gallica*), and cultivar ('Versicolor'). If you make reference to a different species of rose right after that, you'd abbreviate the genus and get, for example, *R. spinosissima* 'Lutea,' the 'Lutea' cultivar of Scotch rose (*Rosa spinosissima*). It could also be called *R. spinosissima* cv. Lutea, with "cv." standing for cultivar. Not every cultivated plant has named cultivars, and even among plants with cultivars, you can often find nonnamed, noncultivar specimens.

When a particular species is not known or it doesn't matter, it gets "sp." in the singular, and "spp." in the plural. So an unidentified but unmistakable member of the genus *Rosa* would be *Rosa* sp. More than one kind would be *Rosa* spp., which is often used to refer to all (or all important) members of a genus. For example, "*Rosa* spp. prefer slightly acidic soil."

Hybrids are indicated by an "x." Harison's Yellow Rose, for example, is *Rosa* x *harisonii*; the "x" tells you that it's a hybrid, which often means that it will not reproduce true to type, if it can reproduce at all. Harison's Yellow Rose is actually a cross between *R. gallica* and *R. spinosissima*—an interspecific hybrid, meaning a cross between two species of the same genus. There are also intergeneric hybrids (hybrids between one genus and another) but they are far more unusual. That would be like a human hybridizing with a chimpanzee (since we're two different genera in the same taxonomic family), or a housecat with a tiger. Freaky.

I've taken some direction from each of these classification concepts in organizing the plants that follow. All of the members of the mustard family (Brassicaceae) are in one place, for example, because they share similar cultural environments and are subject to the same pests. Because of their primary uses in the kitchen, beets appear in "Other Root Crops" and Swiss chard appears in "Other Leafy Greens," respectively, even though they are, in fact, the same species. Also for culinary reasons, Malabar spinach and New Zealand spinach are grouped with plain old spinach, even though they're not at all closely related. It's as much art as science. Maybe more.

Each crop section includes the scientific names for plants, the general cultural conditions that they prefer, and a discussion of any relevant companions (plants that coexist with them to mutual benefit), likely pests, and related plants. I've sometimes included related plants just for interest (okra is second cousin to cacao!) but mainly for reasons of crop health. Related plants often fall prey to the same pests, so it's good to have an idea of the family tree. Eggplants, tomatoes, potatoes, and peppers are all closely related, for example, so planting one crop in a given plot one year and one of its cousins the next doesn't count as rotation. They're all also related to petunias; so, if you find your flowers being gobbled up by something, it might be best to get rid of the petunias before the mystery pest spreads to your heirloom tomatoes. And you should avoid smoking near them, since they are all related to tobacco, and you could end up blowing clouds of contagious tobacco viruses onto them. None of these relations is obvious or intuitive; you just need to study up a little on your crops.

In most places, I've listed a few recommended cultivars for each crop. How did I do so? Well, I checked out recommendations from a cross section of cooperative extensions in climatically diverse parts of the country. If the cooperative extensions covering steamy Alabama, frigid Maine, foggy Washington, and arid Arizona all recommend the same cultivar, for example, it's a good sign that the cultivar is pretty universally tough. Where appropriate, I've tailored the cultivar recommendations to different regions.

I've also indicated where a cultivar is an F_1 hybrid. When you first cross two varieties of a given plant, the resulting offspring (in this case, seeds) often show *hybrid vigor*—they grow bigger, mature more quickly, and yield more. The main disadvantage of F_1 hybrids is that their children (the F_2 generation) lack the vigor,

uniformity, and predictability of the F_1 parents. Some farmers object to F_1 hybrids because it's worthless to save the seeds from one year's crop to plant the next; you need to keep on going to the seed company year after year. Cultivars not listed as F_1 would be considered heirloom under the most accepted standard for the term—they are open-pollinated, meaning that planting their seeds will produce offspring essentially like the parent.

After some discussion of how to choose particular plants, this chapter covers mainly herbaceous crops, most of them annuals. It also includes perennial and biennial plants grown as annuals—in other words, just for one season—either because they will not survive the cold or because they lose productivity after the first year. This category includes some of the best-known, best-loved vegetables, including carrots, peppers, and garlic. (Chapter 7 involves mainly woody crops, most of them perennial.) Most of the crops discussed are clearly suited for everyday urban farming (for example, tomatoes), while some more exotic crops are included for the adventuresome or homesick (such as groundnut or ulluco), and some make the cut simply because even though they are not ideal urban farm crops—I'm looking at you, corn—people will want to grow them anyway.

Your Choice

You can't change your climate, and there's only so much you can do to alter the ground beneath your feet, so for most farmers, plant selection is the main variable over which you have control. You can tweak, but not defeat, Mother Nature. That in itself gives you plenty of room to make choices. T.S. Eliot once said that "freedom is only freedom when it appears against the background of an artificial limitation"—a sentiment surely as applicable to natural limitations. Think of being free to grow what matches your aspirations and restrictions, rather than constrained by what does not. You probably have plenty of choices.

Climate/Cultural Conditions

The first consideration is the prevailing cultural conditions. Choosing the wrong plant for your area can lead to pest infestations, disease, potential crop failure, and almost certain disappointment. I once tried growing all sorts of subtropical fruits, for example—even joining the California Rare Fruit Growers—while living in a dark Bronx apartment. It was not a match made in Heaven. Unless you have access to climate-controlled greenhouses and high-intensity discharge lamps, geography is destiny, at least from an agricultural perspective. Take apples and oranges. About 80 percent of the US commercial apple crop is grown in four states: Washington, New York, Michigan, and Pennsylvania. About 90 percent of the US commercial citrus crop is grown in two states: Florida and California. Notice a pattern? You can grow both in some places, but chances are, only one will really thrive in your area. Read up on the plants you want, and be realistic. After that most basic consideration, choice is based largely on what kind of farmer you primarily are: subsistence, recreational, or entrepreneurial.

Farming Ambitions

What kind of farmer are you, or what kind do you hope to become? The purpose and scale of your farming play a big role in picking appropriate plants.

Subsistence Farmers

For the world's urban subsistence farmers, the main considerations are what they personally eat. "The poorest urban households spend as much as 80 percent of their income on food, up to 30 percent more than is spent by rural families," notes Dr. Gordon Prain, Global Coordinator of Urban Harvest. So, growing the food they would otherwise buy makes a huge improvement in their general economic condition. Most urban farmers in the United States are likely to be better off than the global norm, but similar considerations apply. If you are counting on farming to make a significant dent in your grocery bill, it makes the most sense to focus on those plants you normally eat that you can grow locally and that would typically be

A loose-leaf lettuce cultivar, 'Cracoviensis', growing in a hanging basket. Unusual varieties of high-value crops might be particularly profitable for urban market farmers. This wouldn't be the first cultivar of lettuce I'd choose, however.

most expensive in the marketplace. For example, if your main produce staples are tomatoes, cabbage, and corn in roughly equal quantities (a strange diet, yes, but good for illustration), and you have a plot big enough to grow a season's supply of one of them, then you should choose the one that would be the most expensive and continue to buy the other two.

The secondary consideration for subsistence farmers (or very small-scale entrepreneurial ones) is the prospect of extra growing capacity and what to do with it. Here, it makes sense to focus on the highest yielding, lowest effort crops. In many urban areas, this might be "microgreens," which are essentially just baby versions of standard vegetables such as kale or radishes. They command a premium in many markets and have the added advantage that one can grow them successively for multiple yields throughout the growing season. Another possibility is to grow vegetables and fruits that are not readily available locally. Just make sure that they are not so exotic that no one will buy them.

Recreational Farmers

Beyond physical limitations, recreational farmers are limited only by their own tastes or whims. I, for example, am a hopeless dabbler. While I never seem to produce a great quantity of food, I definitely produce a wide variety. And in trying so many different crops, I am slowly building a list of "winners" (and even more quickly, a list of "losers").

Entrepreneurial Farmers

Entrepreneurial farmers face many of the same marketable-production considerations that subsistence farmers face: deciding which crop—or combination of crops—will reliably command the highest price.

Subsistence and entrepreneurial farmers both often benefit from having a variety of crops, some long-season and some short-season. This hedges against crop failure of any one plant, spreads income across a range of time, and frees the farmer from having to harvest everything at once. Think of it like diversifying an investment portfolio. Crops for market-driven farmers are usually divided between high-value crops (like profitable but risky stocks) and low-value ones (like low-yielding but relatively safe bonds).

Crops typically considered high-value include greens (such as spinach and lettuces), carrots, cucumbers, herbs, radishes, scallions, summer squash (zucchini, pattypan, and so on), tomatoes, ornamental flowers, and mushrooms. Most take a relatively short time to reach a sellable state, and do not stay fresh long, which puts a premium price on them in the market because they often need to reach it quickly and remain refrigerated. Low-value crops are often clunky, take longer to reach a sellable state, and have a longer shelf life. Examples include head cabbage, many beans and peas, potatoes, and winter squashes.

Other Considerations

Urban farmers—especially subsistence and entrepreneurial ones—face the same basic questions encountered by farmers everywhere:

Do you start with seeds or seedlings? Seeds take longer and may require special equipment (such as artificial lighting or a homemade hoop house), but they are cheaper and provide a much wider variety. Purchased seedlings take off more quickly with less care, but they cost more and are subject to local availability. If you have other small-scale urban farmers in your area, a profitable strategy might be to grow your own seedlings—some for your farmstead and the rest to sell to others unwilling to go through the hassle of seed starting.

Should your focus be on annuals or perennials? Many urban farms focus on a single growing season, planting only annuals, or growing biennials and perennials as annuals. Most of the big vegetables fit into this category, and the beauty of it is that you can change your crops every year to meet market demands. Yet there is much to recommend perennial tree and shrub crops, too. They typically have deep roots, improving soil tilth, preventing erosion, and bringing up nutrients. Some, such as mesquite (*Prosopis* spp.), goumi (*Elaeagnus multiflora*), sea buckthorn (*Hippophae* spp.), and black locust (*Robinia pseudoacacia*), restore soil fertility by fixing nitrogen. Perennials are also champions at sequestering carbon, which is increasingly seen as important to the environment. Perennials are a favorite in permaculture, not only because of their functions but also because they are often low maintenance. In fact, a central observation of permaculture is that trees—the ultimate perennials—are the apotheosis of ecological succession. Leave a given plot of land alone, and after anywhere from a decade to millennia, it will become

With a little planning, you can grow enough seedlings to meet all of your needs and sell to others.

a forest. Why fight Mother Nature? That being said, if you have only containers or a short lease, it's not practical to plant a mini-orchard.

Would the premium commanded by organic or biodynamic produce justify the extra expense of certification? And might following the standards without certification yield other, non-monetary benefits, such as the sense of being a producer, a connection with the earth, or a better sense of control over your child's diet? Most of the urban farmers I've spoken with choose to grow organically with varying levels of strictness—none of them primarily for profit motives.

Should I choose to plant native or exotic species? Even within those categories, there are welcome natives and exotics, and unwelcome ones. Most of the major staple crops such as wheat, barley, and rice—not to mention vegetables such as broccoli and fruits such as oranges—have been introduced from abroad within the past few hundred years. Yet no one experiences a problem with rampaging tangerines. American gooseberry, which is native to our continent, is endangered or of special concern in Rhode Island and Illinois but classified as a noxious weed in Michigan. It all depends on your geographical context. It's not just a question of a plant that might potentially run amok; it involves a whole range of considerations, including diseases that the plant might carry. If you live in California, for example, it's unlikely that you could legally import any citrus fruit into the state without sending it for quarantine first, since the state's citrus production is too valuable to risk. Regions with commercially important white pine forests often ban all members of the genus *Ribes* (which includes gooseberries) because they help spread a fungal disease.

If you can buy a particular crop at a small local farmers' market, it's very likely not a problem for you to grow yourself. If it's only available at gourmet markets or not even commercially available locally, it's worth checking with your cooperative extension and/or relevant state agency before you buy it for planting.

The National Resources Conservation Service of the USDA has an excellent online tool to help you sort through many of these considerations and others you never even dreamed of at www.plants.usda.gov/adv_search.html. If, for example, you live in Whitley County, Kentucky, and want to find a dicotyledonous perennial plant that is native to the lower forty-eight states and is found in your county, not invasive or noxious, and a producer of seeds, nuts, or berries—*bang!*—you'd have thirty-nine results, starting with *Amelanchier arborea* (common serviceberry) and ending with *Vitis vulpina* (frost grape).

The biggest category of food crops in terms of both production and consumption has got to be herbaceous plants. They may be annuals, biennials, or even perennials, but the common thread is that they don't produce the pith needed to be woody and the top parts often die down to the ground in winter.

The Exotic Debate

The choice between native or exotic plants—and whether those labels are important or even coherent—is a hotly debated topic within the horticultural community. On the one hand, introduced species can apparently wreak havoc on a new habitat with favorable cultural conditions and no natural enemies, such as kudzu in the Southeast. On the other hand, all land plants started on what was originally just one continent, and seeds are constantly being dispersed by wind, water, and wildlife. Even new volcanic islands soon attract plants and animals. Things move around.

Noted permaculturist Toby Hemenway describes permaculture's view as hierarchical and cautious: "First, use a native to fill the desired role if at all possible. If no natives for that niche exist, then use a tested exotic. Only after a great deal of research would a person then consider a small-scale introduction of a new exotic; and, to be honest, I have never done that, don't personally know anyone who has, and don't recommend it."

J.L. Hudson, a self-described "public-access seed bank" with a strongly educational and activist bent, takes a view on his website (www.jlhudsonseeds.net) opposing such caution. "[T]here is absolutely no biological validity to the concepts of 'native' and 'exotic' species, nor is there evidence that man's introduction of species into new habitats has any negative impact on global biological diversity. On the contrary, the aid we have given species in their movement around the world has served to increase both global and local diversity. It is one of the few human activities which is beneficial to the nonhuman creation."

As at-odds as these opinions may appear, they are premised on a shared observation: intact ecosystems tend to resist non-native "invaders," and disturbed ones invite them. The best defense against invasion, therefore, is to restore and protect the ecosystems we have.

Herbs

What better place to start a discussion of herbaceous plants than herbs themselves? Herbs are usually things that you can recognize when you see them but cannot easily define. What is really the difference between spearmint and arugula, for example? Sure, one is perennial (spearmint) and the other annual, and you'd never dream of making an arugula julep. But they are both herbaceous plants with pungent, edible leaves. Here, I use the term herbs to mean edible plants used mainly as a flavoring rather than as the main event, and used fresh rather than dried (which would be the domain of spices). Herbs are great choices for the urban farmer because they tend to be easy to grow in pots, you can harvest just a bit at a time (as opposed to, say, broccoli), and they are often expensive to buy in stores.

Although there are many varieties of herb, many of the most popular ones can be divided into two groups: dry and moist. Since they have similar cultural requirements, you can companionably combine many dry herbs or many moist herbs in the same containers—just not dry and moist with each other.

Popular dry herbs include rosemary, thyme, oregano, culantro (Mexican coriander, *Eryngium foetidum*), sage, bay, and marjoram. They tend to like very well-drained soil that's neutral to slightly alkaline. Most prefer poor, not particularly fertile, soil.

They're all perennials, and a tough bunch. All can survive with care in terra-cotta pots, which would dry out too quickly for moist herbs. You do need to water them sometimes, however.

Popular moist herbs include basil, mints (*Mentha* spp.), chives, parsley, cilantro/coriander (the "true" coriander, *Coriandrum sativum*), dill, chervil, lemongrass (*Cymbopogon citratus*), rau ram (Vietnamese coriander, *Polygonum odoratum*), and French tarragon. They tend to like slightly moist soil that's neutral to slightly acidic. Most prefer compost-rich, fertile soil. They're a mix of annuals, perennials (tender and otherwise), and biennials.

In addition to these basic culinary herbs, there are a whole bunch more grown for other purposes, including cut fresh and dried flowers; essential oils, potpourri, and aromatherapy; teas and tisanes; bee food; and various medicinal uses. Lavender fulfills all of these purposes, for example, and has been a valuable commodity since ancient times. As with the primarily culinary herbs, many of these other herbs can generally be divided into dry and moist types. Among the dry kinds are lavender, *Artemisia* spp. (which includes tarragon [*A. dracunculus*] as well as various unfortunately named mugworts and wormwoods). Moist types include lemon balm and lemon verbena, both gorgeously scented; scented geraniums (*Pelargonium* spp.); borage; many catnips (*Nepeta* spp.); and bee balms (*Monarda* spp.).

Herbs are not often eaten in quantity, so they may not figure prominently in diet analysis. Many are very high in vitamins, however, as well as in minerals and antioxidants. In fact, culinary herbs are like a who's who of phytonutrient-rich plants. Herbs also play a major role in integrated pest management (IPM). For many commonly grown crops, some herbs are believed to repel harmful insects, to distract or divert them (trap crops), or to attract beneficial insects (insectaries). There are many worthy uses for herbs beyond the decorative sprig on your hamburger.

Eritrean basil (*Ocimum gratissimum*)

Miscellaneous Mustards

When your mother said "eat your vegetables," she was probably referring to those in the family Brassicaceae, the mustard family. It is a ridiculously broad category, dominated by members of the genus *Brassica*. Most hardly resemble the family patriarch in his little yellow jar, except for one characteristic: a sulfury bite. This genus is so big, and so diverse, that it deserves to be considered in a class by itself.

Members of the family Brassicaceae are also sometimes called crucifers because they have four-petal flowers, resembling a cross. Crucifers are a wonderful testament to the power of centuries of human selection, with several species bred for such specific traits that you might not even realize that they are the same species. The species *Brassica oleracea* alone, for example, includes broccoli and cauliflower (bred for flowers); kale, collards, and cabbage (bred for leaves); and kohlrabi (bred for stems)—not to mention Brussels sprouts and other varieties. Similarly, the species *B. rapa* includes snow cabbage, bok choy, turnips, tatsoi, and rapini (broccoli rabe), to name a few. The third heavyweight species, *B. napus*, includes rutabaga and canola (now those are strange bedfellows). Not bad for the tiny mustard seed.

Most crucifers share the same basic cultural requirements, including cool weather, full sun to part shade, and slightly acidic soil with a pH of about 6.0 to 7.0. Soil should be rich in organic matter and preferably amended with compost or composted manure; fertilizing beyond a few weeks after planting is generally not recommended. Mulching is desirable to maintain moist, cool soil. Crop rotation is paramount; err on the side of safety and wait at least a year before planting a second crucifer crop in the same soil. Closely following a particular crop's cultural preferences will make for a healthier plant better able to resist pests and diseases—of which there are many.

A Note on Cultivars

I've tended to list cultivars with broad geographic appeal—ones equally recommended by cooperative extensions in several different regions. That doesn't mean that there isn't a better cultivar for your particular location; there almost certainly is. Check out your local cooperative extension and regional seed banks for recommendations more closely tailored to your city.

In warmer climates, crucifers generally can overwinter, being planted in fall and harvested in spring. They are largely also self-incompatible, meaning individual plants cannot pollinate themselves. Since most are grown to be eaten before they set seed, this raises little concern for the casual gardener. If you hope to save seeds, however, you should isolate the particular cultivar you want reflected in the seeds with a covering to prevent insect pollination from other crucifers or through a *Brassica*-free perimeter of 200 feet (61 m) or more. Remember that this includes weeds in the family Brassicaceae, such as shepherd's purse.

Broccoli and Cauliflower

Broccoli and cauliflower both adhere to classic crucifer cultural requirements. Their aversion to high temperatures means that the best time to plant them varies widely by region: late spring/early summer in the coldest zones, mid-summer in cool zones, late summer in warm

zones, and fall in hotter zones, where they can be grown over winter for a spring harvest. After transplanting, both

take about a month and a half to four or more months to harvest, depending upon the particular variety. 'Premium Crop' (about 55 days from transplant) and 'Early Dividend' (≈45 days), both F_1 hybrids, are standard broccoli recommendations; 'Snow Crown' (also F_1; ≈55 days from transplant) is a cauliflower recommendation. Rapini, or broccoli rabe, is like a loose broccoli and takes about a month and a half to two months. 'Super Rapini' (≈60 days) is a common recommendation.

Kale, Brussels Sprouts, and Collard Greens

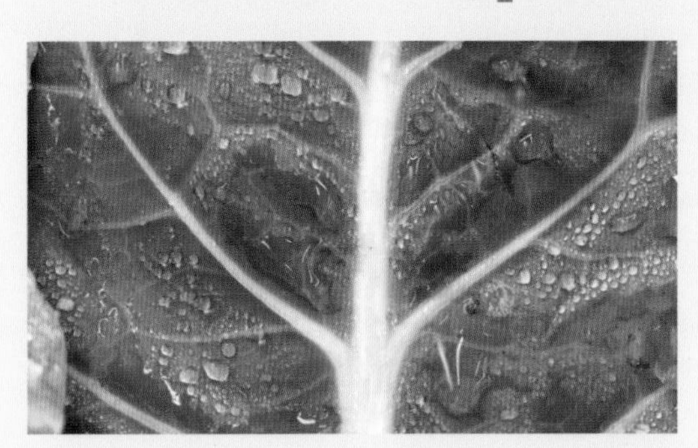

Collard greens

Cabbage

These are among the toughest of the crucifers, being harvestable even under snow. In fact, a light frost is often recommended to sweeten their flavor. Collards are also one of the most drought- and heat-tolerant, although they still favor a cool growing season. In addition to its hardiness and the nutritional value it shares with other crucifers, kale has two great virtues: a short growing season (about 60 days) and some beautiful cultivars resembling giant flowers that can double as ornamentals. 'Dwarf Blue Curled' and 'Red Russian' are commonly recommended, yielding ≈55 days and ≈50 days from transplant, respectively. Baby greens can be harvested in half that time. Brussels sprouts and collard greens both require about three months after transplanting. 'Prince Marvel' (F_1, ≈80 days from transplant) and 'Long Island Improved' (heirloom, ≈90 days from transplant) are recommended Brussels sprouts; 'Georgia' and 'Blue Max' are F_1 hybrids often recommended for collards (both ≈70 days from transplant).

Watercress

Watercress, Garden Cress, and Arugula

Watercress (*Nasturtium officinalis*; confusingly, unrelated to the flower nasturtium), garden cress (*Lepidium sativum*), and arugula (*Eruca vesicaria* var. sativa) are cruciferous contributions to the world of salads. All three enjoy moist conditions—both cresses can grow in water, in fact—although arugula is the best adapted of the three to drier conditions. All appreciate shade, particularly in summer. Unlike most crucifers, watercress prefers a slightly alkaline pH—7.2, to be exact—though it will grow in other conditions. These salad greens grow quickly (giving rise to arugula's other common name, rocket) and can also be enjoyed as sprouts. All are typically sold as the species rather than as any named cultivar. Watercress can often be purchased in a store, rooted in water, and then planted. Garden cress takes about three weeks from germination to be harvested, arugula five weeks, and watercress seven weeks.

Cabbage

It may be the mustard family, and other members may be more in vogue, but surely the two species of cabbage (Western, *B. oleracea*, and Eastern, *B. rapa*) have been the heavy lifters of the family for millennia. In fact, an alternative name for crucifers, "cole crops" is from the German word for cabbage, *kohl*. The ancient Egyptians ate them. Charlemagne ordered them planted. They provide Germans with their sauerkraut, Koreans with baechu

Cabbage

kimchi, Ethiopians with atkilt wot, Hungarians with Töltött Káposzta, Salvadorans with curtido, the Irish with colcannon, American picnickers with cole slaw, and many other peoples with homey, cheap, and nutritious food.

There is an appropriately broad range of cabbage types. Western cabbage (*B. oleracea*) includes both heading varieties (the bowling-ball kind) and nonheading (loose) varieties. The heading types include red, green, and Savoy types, with the latter being crinkly and especially cold hardy. There's a wide range of maturation times, from about two months to more than five, so make sure that you understand the type you're getting. Some are meant to grow in spring for harvest in late spring/early summer, for example, while others grow summer to autumn/winter or even early summer to early the next spring. You can plant different varieties at the same time for near-continuous harvest. F_1 hybrids 'Savoy Ace' and 'Gonzales' produce in 65 and 75 days from transplant, respectively.

China has developed so many varieties of its own cabbage (*B. rapa*) that it's difficult to know which is which, especially because of differences in dialect. Bok choy is particularly troublesome in that different parties use it to describe at least three different varieties of Chinese cabbage. There are heading varieties (often called white cabbage, snow cabbage, Napa cabbage, or pe-tsai) that are looser, whiter, and more oblong than western cabbages. They mature sooner than western varieties, taking about a month and a half to three months. There are also smaller, shapelier varieties with tight, plump, white stalks topped by loose, very green leaves. When they're not called bok choy (which they are around where I live), they are often called flowering white cabbage, pak choi, or choy sum. They take between a month and a month and a half to mature. 'Jade Pagoda' (F_1, ≈65 days from seed) is a recommended cultivar of the larger, headingkind, and 'Joi Choi' (F_1, ≈55 days from seed) of the smaller, nonheading kind.

Whatever the maturation time, you can always harvest baby cabbages or peel off individual leaves as you need them. In addition, heading varieties can often be kept in the ground weeks after they have reached full size.

Radishes, Turnips, and Rutabagas

Part of the mustard family, although not a *Brassica*, the lowly radish (*Raphanus sativus*) is one of the unsung heroes of the vegetable world. The fastest varieties can go from seed to table within four weeks; indeed, the radish genus, *Raphanus*, means "quickly appearing." They can be fiery or mild, tiny marbles or monsters longer and heavier than a toddler ('Sakurajima' daikon radish). Though more valued for its root, the whole plant is edible—leaves, pods, seeds, and all. Many farmers interplant fast-growing radishes with crops such as lettuce to provide shade for the young seedlings, or they use them as "catch crops" grown in extra space to improve the soil, suppress weeds...and provide more radishes. No less a luminary than Masanobu Fukuoka employed them to loosen soils. *Wyman's Gardening Encyclopedia* calls it "the easiest of all vegetables to raise." 'French Breakfast' (oblong red and white, ≈30 days from

Radishes

seed) and 'Snow Belle' (round all-white, ≈30 days from seed) are good bets.

Turnips and rutabagas round out the big trio of crucifer root crops. Although they belong to different species (*B. rapa* and *B. napus*, respectively), they look, taste, and grow similarly. Both the greens and the swollen roots are edible. Turnips are the most heat-tolerant of the bunch and take between one month (for greens) and two-and-a-half months (for roots) to mature from transplants. Rutabagas are the most cold tolerant and take about three months. 'Tokyo Cross' is a popular turnip variety (all-white F1 hybrid, ≈35 days from seed, greens harvestable earlier), and 'American Purple Top' is a popular rutabaga variety (heirloom, ≈90 days from seed).

All three crucifer root crops favor deep, loose soil. Plentiful, consistent moisture helps them grow quickly, stay crisp, and stay mild. Too much heat and too little water make them corky and sharp. All three are also susceptible to deficiencies in boron and manganese, so it might be worth testing your soil if you're counting on a bumper crop of any of them.

Companions, Relatives, and Pests

In addition to those previously listed, many other popular and nutritious vegetables are crucifers, typically with the characteristic mustard bite. Among those most commonly grown are horseradish, mizuna, and canola. Sweet alyssum, stock (*Matthiola* spp.), dame's rocket, and honesty (*Lunaria* spp.) are crucifers grown as ornamentals. Nasturtiums (*Tropaeolum* spp.) and candytuft (*Iberis* spp.) are somewhat more distantly related. Weeds in the mustard family include shepherd's purse (*Capsella bursa-pastoris*) and pepperweed (*Lepidum* spp.).

Crucifers are as popular among pests as they are among people. In fact, crucifers suffer a virtual rogues' gallery of invertebrate pests, including cabbage aphids, maggots, and worms; crickets, earwigs and slugs; cutworms and wireworms; darkling beetles and diamondback moths; weevils and curculios; thrips and whiteflies; all manner of loopers and nematodes; and even pests apparently cheating on their main squeezes, such as beet armyworms and seedcorn maggots. I still don't know what gobbled up the leaves of my crucifers this year, but they strolled right past lettuce leaves to do so.

Crucifers are also susceptible to many diseases, most notably club root, which can take decades to clear once it infects the soil. The University of California's Statewide Integrated Pest Management Program advises liming club-root-affected soils up to a pH of 7.2, since the disease also enjoys acidic soil. (Don't worry, most crucifers will tolerate up to a pH of 7.5).

Where pests such as cabbage worms, cutworms, and cabbage loopers are a problem, you can use floating row covers to protect the crops from infestation. These row covers are like superlight, porous, translucent blankets that can be spread right on top of the plants, without support such as hoops; hence, they are "floating." Already-infested crops can be treated with Bt (*Bacillus thuringiensis*), a bacterium readily available in powdered or liquid form that hits caterpillars or other larvae like anthrax, to which it is, in fact, quite closely related. It apparently does not have the same effect on human health, however; almost two-thirds of corn grown in the United States has been genetically modified with a gene from the bacillium to achieve the same effect.

One would expect the mustard family to go extinct with so many enemies, yet there is no shortage of fluffy mounds of mizuna; broad, frilled leaves of kale; and cabbages that resemble bowling balls in size, shape, and weight. Good cultural practices are key.

Good Stuff from BADSEED

At their plot in suburban Kansas City, Missouri, Brooke Salvaggio and Daniel Heryer had what any aspiring urban farmer would imagine to be a near-guarantor of success: two and a half acres of land. Two and a half! Even with a house on site, it's like forty acres in the country. Yet even the smallest-scale urban farmers could claim two things in common with them: neighbors and zoning laws. And in the case of BADSEED Farm, the former has used the latter to put the kibosh on goatkeeping and other farming endeavors.

At first, of course, urban farm life seemed dreamy to BADSEED founder Salvaggio. Her grandfather owned the property and allowed her to shave off the pristine lawn and start farming. Goats and chickens were introduced originally just to produce eggs, cheese, and milk for the committed homesteading couple. As Salvaggio puts it, "Eating just kale and beets all the time doesn't cut it."

They even developed a four-part rotational grazing system for the livestock. The goats would munch on each section for a week, trimming it down and blanketing it with manure. Then they'd move to the next plot and the chickens would move in, removing any worms or other undesirables from the goats' droppings ("holistic parasite management"), eating stray weed seeds and bugs, and leaving their own excremental souvenirs. By the time the goats had returned to the first plot, a month had passed and there were new, freshly fertilized grasses and forbs. It was healthy and productive for the land, the animals, and the farmers, who opened a BADSEED CSA program.

This was not, however, universally welcome in a leafy suburban neighborhood. Suddenly, they faced a storm of apparent city violations: too many goats, not enough of a setback for the chickens, impermissible use of volunteers (considered "employees") in a home-based business, improper running of a business on-site (because BADSEED Farm CSA members picked up their goods there). Such challenges would crush the hopes of many would-be urban farmers, but Salvaggio and Heryer persisted. They gave up the herd of goats (the entire herd, so they could stay together), centralized the chickens, turned away volunteers, and opened a storefront for CSA members.

Salvaggio and Heryer are also resourceful. Although both have some background in agriculture (she earned her stay around the world volunteering on organic farms through World Wide Opportunities on Organic Farms [WWOOF], and he worked with the Kansas City Center for Urban Agriculture), they also have business savvy. They grew produce not just for their own consumption but for the CSA, for restaurants, and for retail sale at their storefront downtown and elsewhere. He provided a fruit-tree maintenance service; she arranged flowers and baked wedding cakes—all local and organic, of course.

They hosted related events and classes and offered organic gardening consultations. Salvaggio's advice: "Diversify, diversify, diversify."

Alliums: The Glorious Onion Family

Alliums include onions, garlic, shallots, leeks, scallions, chives, and many other tasty relatives, such as ramps, ramsons, and meadowleeks. Most prefer full sun and well-drained, reasonably fertile soil. They are perennial bulbs, though some have been selected primarily for the bulbs (onions and shallots), others primarily for the leaves (leeks, scallions, and chives), and some for both. What you may not realize is that they tend to have attractive, lollipop-like flowers, too. Delicious, perennial, beautiful—what's not to love?

Onions

Onions, Shallots, and Garlic

These may all be grown from seed, but since they may take a long season to plump up, there's an easier way. For onions (*Allium cepa*) and shallots (also usually *A. cepa*), you can buy "sets," or tiny little dormant bulbs grown from seed the year before. With garlic, you can just plant individual cloves (pointy side up and papery, flat, blunt end down). Scallions can be the greens of a regular, bulbing onion; the greens of a non-bulbing onion; or the greens of a closely related species, *A. fistulosum*, usually called Welsh onion.

Onions are divided by daylight requirements into short day (for the South) and long day (for the North). Big, sweet, commonly recommended bulbing onion cultivars are 'Yellow Granex' (≈90 days from transplants; this is the variety used for Vidalia onions) in the South and 'Ailsa Craig

Exhibition' (≈95 days from sets, ≈110 from seed) for the North. 'Evergreen' (≈65 days from seed) is a nonbulbing counterpart to use for scallions. For shallots, 'French Red' is a common recommendation; for garlic, try 'German Extra Hardy.' Maturity times for both vary by region.

Chives

Onion chives (*A. schoenoprasum*) and garlic chives (*A. tuberosum*) are both worth growing from seed. Any variety is worth trying, but typical recommendations include 'Grolau' for onion chives and 'Nira' for garlic chives. Both take about 80 days from seed. You can tell them apart when they're growing because onion chives (usually marketed simply as "chives") have round, tubular leaves and (usually) pink flowers. Garlic chives have flat, straplike leaves and (usually) white flowers.

Chives

Tree Onion

Tree onion

If you're going to pick just one allium, a good choice is the tree onion, also known as the "walking onion." The bulb produces large, round leaves that are edible like scallions; topset bulblets (instead of flowers) that can be eaten like tiny onions; and an underground bulb the size of a shallot. The various names for the variety come from the fact that late in the season, the sets can weigh the stem down to the ground, from which the top sets can begin growing. It's like a vegetable slinky. Although there are cultivars of the crop, they do not appear to be easy to come by. Unnamed varieties should work fine.

Companions, Relatives, and Pests

Exactly where alliums fall within the phylogenentic tree is not universally agreed upon, but they are probably closely related to society garlic (*Tulbaghia violacea*) and asparagus, and anything in the lily family. To err on the side of caution, just assume they're related to anything grown from a bulb. Alliums are pretty tough but can be aided by a range of relatively easy cultural practices. First, plant resistant varieties where particular pests or diseases are expected. Plant only in well-drained soils, preferably in dry weather, and provide regular watering from above if you don't get much rain (to wash away any thrips). Remove competing weeds, which can harbor pests. Rotate alliums out of any given plot for two to four years at a time, preferably with grain crops.

Other Root Crops

Botanists might argue whether a root vegetable is technically a tuber, corm, taproot, tuberous root, hypocotyl, or other structure, but they're all good. They are often colorful, tasty, and high in carbohydrates. They tend to appreciate extra phosphorous and potassium, and they can be sensitive to boron deficiencies. Regular watering is a key factor in their general health.

Carrots and Parsnips

These belong to the same family and have similar cultural requirements. They both prefer deep, friable, cool soil and a long, cool growing season. Carrots want full sun but tolerate a wider range of temperatures (about 40–85 degrees Fahrenheit). Parsnips prefer a narrower range of temperatures (about 60–70 degrees Fahrenheit) but tolerate partial shade or full sun. Both will grow in zones 3–10. Carrots and parsnips are both biennial and will not flower the first year. If left in the ground, however, they will flower in the second year.

Companions, Relatives, and Pests

Carrots and parsnips are umbellifers, members of a family of plants that often have hollow and/or ribbed stems, feathery or fernlike leaves, and five-petaled flowers borne in parasol-like clusters called *umbels* (related to the word "umbrella"). Other umbellifers include celery, fennel, and a host of aromatic herbs, including parsley, dill, cumin, angelica, cilantro, caraway, and lovage. They are also related to wild carrot (Queen Anne's lace) and poison hemlock. Many beneficial insects, such as parasitic wasps and predatory flies, feast on the flowers of umbellifers, so it's good to hav a few around (except poison hemlock, which looks like parsley but is significantly lethal).

To keep carrots and parsnips free of diseases and pests, practice crop rotation (planting them after a *Brassica* crop is a good idea), interplant them with traditional companions like alliums, add compost to the soil, suppress weed competition, and try floating covers where airborne pests are a problem.

Potatoes

Who doesn't love potatoes? Potato is the king of the root crops, accounting, on average, for 61 calories of energy per day for every person on Earth. Although most of us may be familiar with a only a few varieties—maybe Russet Burbank in French fries, Kennebec in potato salad, Purple Peruvian in novelty chips, and Yukon Gold for everything in between—there are, in fact, thousands of varieties in nearly every color of the rainbow. Since cultivars' seeds don't grow true to type, most people use small tubers or tuber cuttings known as "seed potatoes."

A potato plant

Potatoes should have loose, well-drained soil with a pH range of about 5.0 to 6.0. A traditional way to grow them is by

A tomato plant hit by blight

"hilling," or mounding more soil around the stem whenever more than a few inches grow above the soil line. Potatoes can be hilled in deep containers by starting them half-full with soil and just filling ever upward as the plant grows. They are beautifully suited to growing in old tires, as well, since new layers of tire can be added, creating a column containing several layers of potato tubers.

Companions, Relatives, and Pests

Unlike other popular nightshades such as tomatoes and peppers, which are grown from seed, potatoes are usually grown from cuttings—cloned—in order to preserve the qualities of different cultivars. Because this vegetative propagation tends to cause potatoes to collect viruses and other diseases, it is important that you buy "certified" seed potatoes that meet state-governed standards of health. (Most nursery-offered varieties make a poiont of being "certified" seed potatoes.)

It's actually not as much for your protection as for commercial producers. Remember the Irish Potato Famine? The same late blight disease that devastated Ireland in the mid-nineteenth century, for example, hit the US Northwest in 2009, dealing a blow to tomato growers as well. That single disease costs $5 billion a year globally and is just one of many potato ills. In addition to crop rotation and other good standard practices, make sure that potatoes are in a sunny enough place that they will dry (no dripping tree branches, for example), don't smoke near them (seriously), and set out shallow trays of beer if slugs pose a problem (they seek out the beer and drown). You can use floating row covers to control Colorado potato beetle problems in the adult stage and use *Bt* treatments for the larval stage.

Oca, Mashua, and Ulluco

If you're not one for the plain-as-a-potato crops that most people grow, you might give oca, mashua, or ulluco a whirl. They are second only to the potato in the Andes, the native region of all four of these root crops. Oca, mashua, and ulluco are reported to be virtually disease- and pest-free; can be grown separately, together, or even with potatoes; offer fresh culinary tastes; and are very nutritious. As with potatoes, you can grow all of them in containers.

Oca and mashua look like the offspring that might result from the mating of a potato with a caterpillar, while ulluco comes in a dizzying array of sizes, shapes, and colors. They are kind of like the potato's scrappier cousins, except for the fact that none of them is at all closely related. Scrappy they are, however, enduring high altitudes, high winds, wide temperature swings, and water runoff amid shallow, rocky soil. Their one major cultural requirement is a long growing season that heads into short-day period (fall or winter in the United States). That's because tuber formation only begins when daylight falls below a certain number of hours: about nine for oca, twelve for mashua, and nine to thirteen and a half for ulluco. Oca and ulluco should be periodically hilled, like potatoes, and mashua should be protected from drought.

Mashua

They are also quite variable crops. Oca apparently tastes a bit like potato, albeit often a pleasantly sour one because of the presence of oxalic acid, which is also in potatoes, spinach, and rhubarb. Raw mashua has a jicama-like crispness and a peppery bite like radish when raw, but loses the spicy heat when cooked. Ulluco can be somewhat mucilaginous, like okra, and tastes nutty. The greens of both mashua and ulluco are edible, as well, the former being related to nasturtium and the latter to Malabar spinach. Oca leaves are also edible, but should be taken in limited quantities and preferably cooked because of the oxalic acid content. The harvested tubers are usually left in a sunny spot for a few days to mellow the acidity and improve the texture.

Companions, Relatives, and Pests

None of these Andean root crops is closely related to another. Potatoes are closely related to other nightshades (see "The Usual Nightshades," later in this chapter), however, including tomatoes, eggplants, and tobacco. Oca is related to wood sorrels, shamrock, and possibly star fruit (*Averrhoa carambola*). Mashua is related to ornamental nasturtiums (*Tropaeolum* spp.) and the canary creeper vine (*T. peregrinum*). Ulloco is closely related to Malabar spinach and more distantly to amaranth, spinach, beets, and rhubarb.

Groundnut

Pre-Columbian North America may not have had the diversity of tuber crops that South America did, but the tuber it did have was a good one: the groundnut, *Apios americana*. Untrained eyes may mistake the groundnut for a weedy vine, never realizing the bounty underground: tubers, sometimes as large as potatoes, with about three times the protein. Other sources agree that the tubers may be as large as potatoes, but only if we're talking about very, very small potatoes. In any case, cultivars developed by Louisiana State University are said to be the best for food purposes.

Even without great masses of tuberish delight, however, groundnut is a useful plant. It fixes nitrogen in the soil, provides edible tubers and seeds, and even boasts pretty, violet-scented flowers. It is a true multipurpose plant.

Companions, Relatives, and Pests

There are two less-documented relatives of groundnut: *A. priceana* (traveler's delight), which is found in just five states centered around Tennessee, and *A. fortunei*, native to parts of East Asia. Groundnut is related to all legumes, including beans, peas, peanuts, and sweet peas (*Lathyrus* spp.). Not widely cultivated for food, groundnuts have few serious pests or diseases. They prefer moist, acidic situations like many *Vaccinium* spp. and, in fact, grow as weeds in commercial cranberry bogs. What is a problem for cranberry growers (unintentional vegetative propagation) would likely not trouble the small-scale farmer, particularly in containers, making for a potential companion to blueberries and other *Vaccinium* spp. with unusual cultural requirements.

Jerusalem Artichokes (Sunchokes)

Despite their name, Jersusalem artichokes (or sunchokes) are actually sunflowers from North America, much more closely related to other sunflowers (same genus) than to real (globe) artichokes (same family). They can—and perhaps should—be grown in large containers; most experts suggest that if you plant them in the ground, you'll be surrendering that patch to them forever. Although the sunchokes have small yellow flowers reminiscent of their giant cousins grown for seed, sunchokes' tubers are their gardening *raison d'etre*. They are usually described as being potato-like but tasting like water chestnuts. They need loose, well-drained soil (they can rot if kept wet) and a three- to four-month growing season. The tubers should survive underground even after hard freezes kill the stem and leaves.

One quirky characteristic that Jerusalem artichokes do share with globe artichokes is that they store most of their carbohydrates as inulin rather than as starch. This makes them, when fresh, less caloric than starchy alternatives (such as potatoes) and less likely to raise blood-sugar levels much.

Companions, Relatives, and Pests

The sunchoke is related to the sunflower and salsify, and all three to a wide variety of weeds, including lawn burweed, common cocklebur, cudweeds, dandelion, English daisy, fleabane, groundsel, prickly lettuce, Canada thistle, and yarrow. It is more distantly related to lettuce. Sunchoke generally grows untroubled by pests or diseases.

Beets

If you're not already acquainted with beets, I suggest you get to know them. Beets are cool-weather crops that prefer a pH of about 6.5 to 7.0 and, like most root vegetables, favor loose, deep soil high in organic matter. They benefit from regular watering and are more tolerant of salt than many other crops. One of the wonderful things about beets—besides the fact that they're delicious and supernutritious—is that they're a two-for-one deal: the fleshy taproot is the headliner, but the leaves are basically Swiss chard (which is the same species selected for greater leaf growth). Two cultivars to try are 'Ruby Queen' (≈60 days) and 'Albina Vereduna' (≈60 days), which is a mild white variety suitable for people worrying about staining everything with a "blood turnip."

Companions, Relatives, and Pests

Beets are related to spinach, chard, orache, amaranth, quinoa, and a number of major common weeds of the goosefoot family, including pigweeds, arrowheads, Russian thistle, and lamb's quarters. You can maximize beets' resistance to diseases and pests by using mulch to supporess weeds, thinning plants sufficiently to allow air circulation, rotating crops, and using floating covers if flying pests abound.

The Usual Nightshades

Some of our most popular vegetable crops are nightshades—members of the family Solanaceae, especially the genus *Solanum*—including tomatoes, peppers, eggplants, and potatoes. Nightshades are usually distinguishable by a number of features, including hairy stems, a shrubby growth habit, alternate leaves, five-pointed flowers that grow in clusters called *corymbs*, seedy fruit, and a distinctive musky smell. Many are extremely poisonous in whole or in part. It is surely for this reason that the English-speaking world refused to eat tomatoes long after the Spanish had introduced them to the rest of the world from South America. Most of the popular yummy nightshades are from the New World (except for eggplants), while the Old World was stuck with ones like deadly nightshade (*Atropa belladonna*), a source of the toxin atropine.

In a strangely appropriate episode of *The Simpsons*, lovable oaf Homer accidentally crosses tomato with tobacco—also a nightshade—producing the revolting but addictive "tomacco" fruit. ("This tastes like Grandma," young Ralph Wiggum says, spitting it out in disgust. "I want more.") You can, in fact, hybridize the two plants to make a tomato containing nicotine, but I wouldn't recommend it. Nor would I suggest following the brilliant season-extending idea of one farmer in Tennessee, who grafted tomato onto a hardier nightshade, jimsonweed (*Datura stramonium*). It worked. Unfortunately, in addition to the longer season, it delivered atropine and the hallucinogen scopolamine. He and his family survived eating the first tomato, but just barely.

Tomatoes

Leaving aside toxic grafts, it would just seem wrong not to grow tomatoes. Tomatoes (which are, botanically speaking, fruits) like warm, well-drained, well-fertilized, moist soil with a pH of 5.5 to 6.5. As the explosion of heirloom tomato availability over the past decade proves, tomatoes come in a wide variety of shapes, colors, and sizes. There are beefsteak varieties preferred for slicing fresh, plum or paste varieties for sauces and canning, and cherry and grape tomatoes for salads and snacking. It would be unthinkable for any self-respecting nursery or seed source

There are several "wild" tomatoes or other species within the same genus. This tiny species, *Solanum pimpinelli-folium*, may even be the progenitor of all cultivated tomatoes.

to offer fewer than five varieties; some offer dozens. Transplants produce fruit anywhere from just shy of two months to nearly three; larger varieties take longer.

The two most immediate considerations, however, are whether the variety is *determinate* or *indeterminate*, and how long it takes to produce fruit. Determinate tomatoes grow with a well-defined central stem, are relatively compact, and produce most of their fruit at once. Indeterminate types get bigger and messier but usually produce more fruit over an extended period of time. Sellers generally tell you the cultivar name, the number of days until the first harvest (after transplanting), and whether the variety is determinate or indeterminate. For container-based plantings without a lot of room, smaller determinate varieties are probably the way to go.

The University of Florida developed the 'Micro-Tom' cultivar (≈80 days from seed), purportedly the world's shortest, which grows no taller than 8 inches (20 cm) and can be grown in a 1-quart (1-liter) container, or three to a typical hanging basket. Ditto for the university's 'Micro-Gold' cultivar (≈85 days from seed). Another commonly recommended, very small variety is 'Yellow Canary' (about 90 days from seed). Slightly larger cultivars for 1- to 3-gallon containers (roughly 4 to 11 liters) include 'Window Box Roma' (≈70 days from transplant), 'Tiny Tim' (≈45 days from transplant), 'Small Fry' (≈65 days from transplant), and 'Tumblin' Tom' (for hanging baskets, ≈65 days from transplant). You also can grow tomatillos (small, usually greenish cousins of tomato that form inside a papery husk) like medium-sized indeterminate tomatoes, with which they share similar requirements.

If you have more room and containers of 5 gallons (19 liters) or more—or raised beds or the ground—you're limited only by cultural conditions. There are far too many wonderful cultivars to list. Be sure you plant early if you have late varieties in cold climates—either that, or acquire a taste for green tomatoes. Consider a mix of cultivars, and possibly a mix of determinate and indeterminate varieties, to ensure a near-continuous bounty throughout your growing season. All tomatoes are warm-season, heavy-feeding fruits that like regular deep watering.

Peppers and Eggplants

Peppers and eggplants are warm-season vegetables but need longer, warmer seasons than tomatoes. Both need about two to three warm months, with nighttime temperatures not falling below 55 degrees Fahrenheit (13 degrees Celsius) for peppers and 70 degrees Fahrenheit (21 degrees Celsius) for eggplants. This makes container culture particularly attractive in colder areas, where you may be able to extend the growing season by taking the potted plants in at night.

Commonly recommended cultivars of eggplants for containers include 'Bambino' (about 45 days), 'Fairy Tale' (≈50 days), 'Ichiban' (≈60 days), and 'Dusky' (≈60 days). All are F_1 hybrids, meaning that the seeds they produce will not produce plants true to the cultivar. Suggested nonhybrids include 'Morden Midget' (≈65 days) and 'Black Beauty' (≈80 days). Commonly recommended peppers for container culture include 'Sweet Banana' (sweet, ≈70 days), 'Jalapeno' (hot, ≈70 days), 'Yolo Wonder' (sweet, ≈75 days), and 'Habanero' (hot, ≈95 days). Peppers like a pH of about 5.5 to 6.5 (the same as tomatoes) and eggplants like about 6.0 to 7.0.

Companions, Relatives, and Pests

All of these nightshades are related to each other as well as to other important crops such as tomatillo, tamarillo, Cossack pineapple and ground cherry, and wonderberry and garden huckleberry. Related weeds include jimsonweed, thornapple, and most anything with "nightshade" in its name. In addition to crop rotation, good ways to protect nightshades include regular ground-level watering and good air circulation (both to keep the leaves dry), organic mulches, early application of dolomitic lime (to prevent blossom-end rot), and prompt removal of any diseased or dead foliage.

Zukes and Cukes: The Cucurbits

While perhaps not as stunningly diverse as the Brassicaceae, or as much the culinary darlings as the Solanaceae, the Cucurbitaceae are certainly no slouches as families go. In fact, historians believe that the very first greenhouse was constructed to sate the Roman Emperor Tiberius's year-round appetite for cucumbers. Cucurbits range in size from cocktail-size cornichons to pumpkins weighing over half a ton, and the family also includes summer and winter squashes, melons, and gourds in addition to cucumbers. They are, for the most part, sprawling annual vines with tendrils.

Hailing from tropical regions, they resent frost as a rule, so you may need to start seeds indoors to be able to grow long-seasoned varieties. One nice characteristic of many cucurbits is that they tolerate partial shade, which is anathema to most nightshades and other popular vegetables.

Cucumbers

Cucumbers (*Cucumis sativus*) are forgiving of soils but generally prefer moisture-retentive (not wet) soils high in organic matter. They must be watered regularly. Cucumbers take about two to three months from planting to reach maximum size. They are usually divided into little cucumbers (including gherkins and

Squash blossoms are both attractive and edible.

cornichons) suitable for pickling and "slicing" varieties best eaten fresh. (You can use slicing varieties for pickling and pickling varieties for slicing, but those are not their best uses).

Bush varieties are available for those with limited space. So-called "burpless" varieties are cultivars bred to be especially mild and easy to digest. Commonly recommended cultivars include: 'Salad Bush' (bush variety, ≈60 days), 'Straight Eight' (slicing variety, ≈60 days), 'Sweet Success' (burpless slicing variety, ≈55 days), 'Marketmore' varieties (slicing, 60 to 75 days depending on version), 'County Fair' (pickling variety, ≈55 days), and the mellifluously named 'H-19 Little Leaf' (pickling variety, ≈70 days).

Summer, Winter, and Pumpkin Squashes

These involve three different species (*C. pepo*, *C. maxima*, and *C. moschata*) that all require cultural conditions similar to those required by cucumbers. Summer squashes are generally thin-skinned, poor keepers, and eaten raw or cooked, while winter squashes are thick-skinned, good keepers, and most always cooked. You can harvest any of them while they're still small and tender rather than waiting until all the days up to harvest have been crossed off and the squash is maximum size.

Summer squashes such as zucchini can be among the most rewarding for urban farmers in that they tend to be fast growing and productive. 'Black Beauty' (≈50 days) and 'Gold Rush' (≈50 days) are commonly recommended zucchini-type summer squashes, and 'Sunburst' ≈50 days) is a pattypan type.

Commonly recommended winter varieties include 'Buttercup' (≈110 days) and 'Sweet Mama' (≈90 days), but the next one I'm going to try is definitely 'Pink

Banana' (≈105 days), which is a giant, banana-shaped winter squash highly recommended by Brooke Salvaggio of Urbavore Farm in Kansas City. Though Pink Banana is difficult to grow, according to Salvaggio, the texture is "like butter." She's a fan of heirloom winter squash in general, the giant pumpkiny kinds, a single one of which often provides enough for "three pies, a big pot of soup, and then…some left over for pumpkin jam or pumpkin butter." Among pumpkins proper, 'Autumn Gold' (≈90 days) and ' Spirit' (≈100 days) are good medium-sized, all-purpose hybrids.

An immature melon

Melons

Unlike many cucurbits, melons need full sun. Honeydew, cantaloupe, and muskmelon cultivars all derive from a single species (*C. melo*), while watermelons are another species entirely (*C. lanatus*). Melons tend to take a long time to mature. Even though urban farmers may have an extended season because of wind protection and the heat island effect, it's probably worth staggering plantings of early-ripening varieties rather than banking on long-season ones. 'Sugar Bush' (≈80 days) is a compact and relatively fast-maturing watermelon, 'Athena' is a muskmelon/cantaloupe type (≈80 days), and 'Earlidew' (≈80 days) is a honeydew-type melon.

Companions, Relatives, and Pests

Melons, squashes, and cucumbers are all related to each other as well as to a variety of obviously named weeks, including wild cucumber and ivy gourd. Having lost the five or so cucurbits I planted this past season, I was consoled to learn that cucurbits, like the brassicas, are poster crops for pests and disease. Recommendations for over a dozen applications of pesticide are not out of the ordinary. For low to no use of commercial pesticides, recommended practices include using floating row covers (except when the flowers need pollinating), growing from transplants rather than seeds, keeping cucurbit plantings away from overhanging foliage that might drip on them, and planting trap crops and insectary plants—a wide variety of each. Sometimes farmers use pheromone traps to find out what pests are in town. And, of course, crop rotation is critical and vigilance is key.

Legumes

Green beans

Legumes are members of the pea family, Leguminosae (sometimes known as Fabaceae).

Whereas the Old World had a near-monopoly on edible crucifers and the New World on nightshades, nearly everyone could claim local legumes since antiquity: soybeans and mung beans in East Asia; peas, chickpeas, and lentils across West Asia and Europe; fava beans, tamarind, and jugo beans in Africa; common beans, peanuts, and groundnuts (*Apios* spp.) in the Americas; and various acacia species throughout Australia.

Without the high-protein, soil-enriching legumes, including beans that could be dried and stored for decades, who knows if humans would have made it? There are many leguminous plants worth exploring, including some shrubs and trees in chapter 7, but the place for most farmers to start would be peas and beans. Though peas and beans are naturally viny, there are many smaller bush varieties available (for runner beans, they're often called "half-runner" beans). Even small cultivars can benefit from some support, however, and larger ones definitely need support—a trellis, a tepee-like arrangement of stakes, or a traditional beanpole. Bush-type beans usually yield sooner but fewer.

Many peas and beans offer multiple crop options: baby shoots, greens, flowers, snap-size pods, or shell-size seeds. Johnny's Selected Seeds, for example, lists a 'Dwarf Grey Sugar Pea' that takes about 32 days for pea greens, about 39 days for blossoms, and about 57 days for snow pea pods. Legume crops can benefit from a treatment of an inoculant containing rhizobial bacteria; the inoculant costs a few dollars. Treated seeds are coated with a fungicide to stop them from rotting in cool, wet weather; untreated seeds are not. Also, legumes do not need much supplemental nitrogen because rhizobial bacteria help them take it from the air.

Fava bean plant

Peas, Fava Beans, and Runner Beans

Like the crucifers, peas (including field, snap, and snow), fava (or broad) beans, and runner beans are all cool-weather crops. This makes them somewhat complementary to nightshades and other hot-weather crops. They like well-drained soil, full to part sun, lots of organic matter, and plentiful moisture after they develop a good root network.

'Dwarf Grey Sugar Pea' (≈60 days) is a commonly recommended snow pea variety, and 'Green Arrow' (≈70 days) is a recommendation for shelling peas. 'Windsor' or 'Broad Windsor' (bush type, ≈80 days) is a commonly available fava bean cultivar. 'Scarlet Runner' or 'Scarlet Emperor' (≈70 days to snap stage, ≈115 days to dry beans) are standard runner bean recommendations, and 'White Half Runner' (bush type; ≈60 days to snap stage) is a common half-runner cultivar. In general, 'Scarlet Runner' is better in the North, and 'White Half Runner' is better in the South.

"Common" Beans

"Common" beans—which include string beans as well as kidney, navy, cranberry, pinto, and other beans—belong to a single species, *Phaseolus vulgaris*. Lima beans are a separate species (*P. lunatus*) but with similar cultural requirements. Common beans and limas are more tolerant of heat than the others; in fact, the soil should be at least 60 degrees Fahreneheit (15.5 degrees Celsius) before you even plant seed or set out transplants. There are too many varieties and purposes for growing them to provide many recommendations, but here are just two: 'Bush Blue Lake 274' is a bush-type snap (or string) bean (≈60 days) and 'Henderson Bush' is a reliable bush-type lima bean (≈70 days).

Cranberry beans

Companions, Relatives, and Pests

In addition to those species previously mentioned, legumes are related to alfalfa, lupines, peanuts, clover, vetch, and many plants you can think of that have a seed pod (such as black locust trees). Legumes are not as troubled by pests and diseases as some other vegetable-crop families. In addition to regular crop rotation, good cultural practices include limiting nitrogen fertilizers, setting out beer traps where slugs or snails are a problem, and watering early in the day.

Leafy Greens

No list of vegetables would be complete without the nutritional powerhouses known as the "leafy greens." True spinach and its imitators are prime examples.

Spinach

Spinach is the chief noncruciferous leafy vegetable, with its prominence evidenced by imitators such as New Zealand spinach, Malabar spinach, and mountain spinach (or orache). The best spinach for you depends upon where you live: I'd guess true spinach straight across the northern states, New Zealand in the Southwest, Malabar in the lower southeastern states, and mountain everywhere else. Little did Popeye realize that the original spinach (*Spinacea olearacea*) might be the wimpiest of the bunch. The real-deal spinach is a cool-weather crop, far more tolerant of cold than of heat. It likes fertile soil rich in organic matter—and regular waterings—but is not very picky otherwise. Spinach can tolerate light shade and should have some in hotter conditions. 'Space' is an F_1 hybrid suitable for spring, summer, or fall planting (≈45 days), while 'Bloomsdale Long Standing' is a crumpled-leaf heirloom best planted in early spring or late fall for overwintering (≈50 days).

New Zealand Spinach

New Zealand spinach (*Tetragonia tetragonoides*) is not closely related to true spinach, though it has a similar taste and is likewise very nutritious. It is tougher than true spinach, liking full sun, tolerating drought and salt, and shedding insect pests like raindrops. It is

less tolerant of cold than real spinach, plus it's a great option for spinach lovers in hot cities—Erik Knutzen and Kelly Coyne, authors of *The Urban Homestead*, have the clambering groundcover all over their front garden in Los Angeles.

Malabar Spinach

Malabar spinach (*Basella alba* [green] or *B. rubrus* [red-stemmed]) is yet another nutritious understudy, not closely related to the other two. It is a climber that likes full sun, hot temperatures (preferably 80 degrees Fahrenheit [27 degrees Celsius]or hotter), and plentiful moisture.

Mountain Spinach

The fourth contender, mountain spinach (or orache [*Atriplex hortensis*]), is actually related to true spinach and is another cool-weather crop. It is more heat tolerant than true spinach, however—and salt tolerant to boot—and can reach 4 or more feet (1.25 m) high.

Swiss Chard

Swiss chard (also known as beet spinach) is actually the same species as beet but has been selectively bred for striking leaves and ribs rather than a fleshy root. Like orache, it is sometimes grown as a more heat-tolerant alternative to spinach. It is also more cold tolerant. Though it adapts to a range of pH levels near neutral, it prefers slightly alkaline soil. An early spring planting will give a long harvesting period. 'Fordhook Giant' appears to be a near-universal, all-green recommendation (≈25 days for baby greens, ≈50 days for full maturity), while 'Ruby Red' ('Rhubarb Chard', ≈30 days for baby greens, ≈55 days for full maturity) takes the cake for red-ribbed varieties. Just make sure that it's really Swiss chard (*Beta vulgaris*) and not the plant rhubarb (*Rheum* x *cultorum* or *R.* x *hybridum*), which looks similar but has poisonous leaves.

Companions, Relatives, and Pests

Spinach, chard, and orache are related to beets and to each other as well as to amaranth, quinoa, and a number of major common weeds of the goosefoot family, including pigweeds, arrowheads, Russian thistle, and lamb's quarters. Disease-resistant varieties are helpful, as is prompt removal of any pest-affected leaf. New Zealand spinach is generally pest free. Malabar spinach should be thinned regularly to discourage fungal diseases, and it is utterly vulnerable to frost. It is related to ulluco.

Lettuce

Lettuces, like true spinach, enjoy cool, fertile, moist, mostly sunny conditions but also a near-neutral pH. Growing lettuces can be rewarding in cooler areas (or in winter in warm areas) because there is a wide variety of colorful, delicious cultivars too fragile to make it to market. Strangely enough, the least rewarding lettuce for urban farmers is likely the one we're most familiar with—the oppressive 'Iceberg' cultivar—because it has very demanding requirements. There are three main classes of lettuce to choose from: the loosely packed butterhead (bibb or Boston) kinds that resemble flowers, the stiff upright romaine (cos) kinds, and loose-leaf lettuces. ('Iceberg' is in another class, crisphead, which is not as easily grown).

As with most vegetables, I would generally prefer staggered plantings of early-type cultivars rather than count on a mass of late-type cultivars. Three commonly recommended varieties that fit this criteria are 'Buttercrunch' (butterhead, ≈60 days); 'Little Gem' (romaine, ≈60 days); and 'Black Seeded Simpson' (loose, ≈45 days). All can be picked earlier as "baby" lettuces.

Companions, Relatives, and Pests

Lettuce is related to chicories (including endive and radicchio) and scorzonera, and more distantly related to sunflowers (including Jerusalem artichoke), asters, and daisies. In addition to crop rotation, good practices include using raised beds, watering regularly, sinking cardboard tubes (such as halves of toilet-paper roll tubes) around young seedlings, using *Bt* to control larvae and floating row covers to control adult predators, and neutralizing slugs and snails using beer traps.

Miscellaneous

As much as the foregoing categories give short shrift to the tremendous variety in each, the "miscellaneous" bin leaves out even more. There are countless crops that might tempt a budding urban farmer, but at least five are worth a special mention.

Corn

Corn is, perhaps, the crop most associated with America (as in North). Popped and buttered, reengineered and syrupified, beloved and reviled, corn holds a special place in our national psyche. The amount of corn that the average American eats in a year weighs about the same as a baby grand piano. It's so cheap and plentiful that it would be a shame to grow it, but maybe that's exactly why you want to see what goes into its production.

Corn likes deep, fertile, well-watered soil in full sun. It shouldn't be planted until the soil is about 60 degrees Fahrenheit, and it takes about 65 to 85 days to mature, so plan carefully. You should plant at least sixteen plants of any one cultivar (four rows of four) to ensure pollination, and the stalks are big: from about 5 to 10 feet (1.5 to 3 m) tall (so you might not want to plant it in front of a window). Also be aware that each plant will produce only one or two ears if all goes well, so don't expect to open up a tortilla factory. Two early- to mid-season hybrid cultivars are 'Earlivee' (yellow, ≈70 days) and 'Peaches & Cream' (bicolor, ≈80 days). A nonhybrid mid-season cultivar is 'Golden Bantam' (yellow, ≈80 days). All of these varieties are 7 feet (about 2 m) tall or shorter, with 'Earlivee' being the shortest, at about 5 feet (1.5 m).

Companions, Relatives, and Pests

Corn is related to other major grains, such as wheat, millet, and oats, as well as to most grasses and sedges. In addition to crop rotation, good cultural practices include regular fertilization and watering during growth, application of *Bt*, and placing a rubber band around each cob top when silks first appear to shut out corn earworms.

Okra

There's little chance that okra is overplanted, although there is much to recommend it, including a short season (making it doable in the North), love of heat (making it great in the South), and preference for alkaline soils (making it unusually well suited for the West). It benefits from high-potassium fertilizers and needs full sun. 'Clemson Spineless' (≈65 days) is a traditional favorite that grows only about a yard high, while 'Star of David' (≈70 days) is an extra-plump cultivar that grows up to 10 feet high, just in case you have an especially expansive farmstead.

Companions, Relatives, and Pests

Okra is related to cotton, hibiscus, and cacao among commercial crops. It is sometimes recommended as a crop to grow with melons and cucumbers because it likes similar cultural conditions. It should be rotated with other crops, of course, and benefits from regular fertilizing, pruning of old pods (wear gloves), and occasional blasts from a hose if aphids are troublesome.

Celery

Celery is a cool-weather crop favoring soil high in organic matter, constant moisture, and a long season. As with 'Iceberg' lettuce, celery is one of those vegetables whose ubiquity suggests ease of cultivation when, in fact, it's somewhat challenging because of its preference for evenly moist soil and a long, cool growing season. 'Golden Self Blanching' (≈85 days from transplant) matures quickly in celery terms, though 'Utah 52–70' (≈120 days from transplant) and its kin are generally considered more resilient.

Companions, Relatives, and Pests

Celery is the same species as celeriac and smallage and is related to other umbellifers, weed and otherwise. It is not a usual victim of pests and diseases. Regular watering is important, as is destruction of any apparently diseased foliage, and you should certainly rotate it with other crops.

Asparagus

Asparagus is an investment. It may take years to produce reliably but will last decades. You don't want to plant it just any old place. It likes full to partial sun; deep, well-drained soil chock full of organic matter; and usually a winter freeze. It tolerates saline conditions, as well. Asparagus plants are sexed—there are males and females, but the males produce more spears because they do not need to produce seeds. As a result, the commercially available varieties are nearly all male hybrids. Various 'Jersey' cultivars ('Jersey Prince,' 'Jersey Knight,' and so on) are well suited to cities with cold winters, while 'UC 157' is better for milder climates. Asparagus will not yield harvestably until the second or third year after planting.

Rhubarb

If one has to cut off a discussion of herbaceous vegetable crops somewhere, it might as well be with rhubarb. It has a whiff of Jekyll and Hyde romance to it, with its delicious red stems and toxic green leaves. Rhubarb is a long-lived perennial, like asparagus, and likes the same cultural conditions (except for the saline part). Both asparagus (often hybrids) and rhubarb (always hybrids) are usually propagated by "crowns," which are starfish-like root sections. Good cultivars include 'Canada Red' (very red, sweet stalks) and 'Early Victoria' (early producing but with greener stalks). Rhubarb should not be harvested until the second year and should not be harvested heavily until the third and subsequent years.

Companions, Relatives, and Pests

Asparagus is a lot like the alliums: a little taxonomically vague but probably related to them and the lily family. Rhubarb is related to buckwheat, of all things, as well as to a wide variety of useful plants and weeds in the Polygonaceae family, including *Coccoloba* spp. (seagrape, pigeon plum, and so on), knotweed, smartweed, *Rumex* spp. (docks and sorrels), and various bindweeds. Asparagus and rhubarb can be grown near each other as well as with horseradish (*Armoracia rusticana* or *Cochlearia armoracia*), another cold-tolerant, sun-loving perennial. Container culture is best for horseradish unless you know you absolutely love the stuff because it can become invasive when put into the ground. (Horseradish is a crucifer.)

Because asparagus and rhubarb are perennial, crop rotation is not an effective strategy. This is all the more reason, however, to make sure that you are using a clean patch of soil and getting disease-free plants (especially for asparagus). Other than that, the best practices are weeding regularly and removing any disease-affected foliage quickly.

Alternative Crops

Not every plant grown by urban farmers is for food. Chrysanthemums for the retail trade, for example, are commonly grown around Hanoi. You don't need to be in subtropical zones to raise flowers, either. Brooke Salvaggio of Urbavore in Kansas City, for instance, grows flowers to use in organic flower arrangements. It might be hard to break into the market for long-lasting, internationally shipped crops, such as tulips and roses, but the urban farmer is at a clear advantage in growing native flowers and those with a short period of bloom—or those that bloom at the "right" time for your region. Cherry, crabapple, pussy willow, forsythia, flowering quince, witch hazel, lilac, and many other shrubs provide whole branches of decorative blossoms. (Even if you have no intention of eating their fruit, many provide food for wildlife, too.) Ferns, hemlock, baby's breath (*Gypsophila* spp.), and other plants provide feathery backgrounds in floral arrangements. Cattail, pussywillow, lotus, Chinese lantern (*Physalis alkekengi*), various vines, and many other plants with interesting pods or other structures make excellent dried specimens.

Luffa—before becoming a bath accessory.

For anyone with the know-how and resources to do so, growing bonsai or orchids (both demanding) is a possibility. There's a seasonal demand for ornamental gourds and pumpkins, and a year-round demand for the related luffa (or "loofah") plant (*Luffa* spp.), which can be made into scrubby sponges of the sort you'd find in a high-end bath store. Speaking of which, it's likely that at least a few of the plants—such as lavender, vetiver, linden, and pine—used to provide fragrance to handmade soaps can grow in your region. Plants used in traditional medicine are another option, and some cities already have zoning laws in place for growing medical marijuana. Many plants, including ornamentals (such as Russian sage or zinnia) and plants with edible parts (including chamomile, saffron, and purple basil), can also be used to make natural dyes.

Dried gourds used as birdhouses

Gourds

Gourds are usually ornamental or utilitarian—like birdhouse gourds that make great homes for birds, or bottle gourds that hold liquid, and all sorts of gourds used to make masks and instruments. Gourds are annuals that usually take ninety to one hundred days to mature, and they are likely good options if you're concerned about soil contamination and want crops that are decorative rather than edible.

Mushrooms

Just as not all plants grown by urban farmers are for food, not everything grown for food is a plant. Sure, there are animals (see chapter 8), but there are also fungi, primarily mushrooms. In fact, the dank, low-light urban conditions that are the kiss of death for most plants are perfect for the cultivation of mushrooms. Mushroom kits are readily available for a wide variety of mushrooms, including reishi, oyster, button, and enoki. Some of the kits consist of a special mushroom substrate inoculated with mushroom spawn, while many others consist of short inoculated logs. The most popular logs are for shiitake mushrooms and can, if cared for, bear mushrooms regularly for years. You can display them as centerpieces on your dining room table, tuck them underneath the kitchen sink, or cloister them in the basement or just about anywhere else within a fairly generous temperature range (about 55 to 80 degrees Fahrenheit [13 to 27 degrees Celsius] for shiitake; other mushrooms vary).

If you really want to get into it, you should know a few things about mushroom biology. First of all, the little capped fellow we call a mushroom is really not the living entity proper; it's just the "fruiting body" of the thing. The real fungal critter is a whiskery, threadlike structure called the *mycelium*. The mycelium grows in a substrate such as soil or a dead log, spreading around to occasionally enormous sizes. Imagine roots on steroids, without any other part of the plant—that is, until it's ready to reproduce. It then pops its fruiting structures out and voila: *champignons*.

As mushroom expert David W. Fischer points out, picking a mushroom is more akin to picking an apple than to harvesting a whole plant. Mushrooms are the organisms' means of producing spores. Large-scale growers order mushroom spawn (which is a spore-inoculated matter similar to grains) from reliable laboratories and then inoculate their own logs (or other substrate), usually by perforating them with a drill and then packing the holes with spawn.

Besides extending your farmstead to include conditions that are less than ideal for plants, there are several other advantages to mushrooms. Many can be harvested year-round and command high prices in places (such as cities!) where gourmets abound. Those that cannot be sold fresh can often be dried and sold later. (In Asia, in fact, that's the main way of using shiitake mushrooms.) From a larger ecological perspective, mushrooms are great because they are the ultimate resource misers, requiring little if any light, supplemental heating or cooling, or fertilizer.

Another option is to gather mushrooms from the urban wild, though I wouldn't recommend it unless you really know your mushrooms. I mean *really* know them—like *PhD in mycology* know them—especially if you plan on selling them. With names like death cap (various *Amanita* spp.), destroying angel (other *Amanita* spp.), and deadly dapperling (*Lepiota brunneoincarnata*), you can't say you haven't been warned that mushrooms can be dangerous. That's truth in advertising. The question is: can you tell one product from another? If you do have that PhD and plan on gathering wild mushrooms to sell, extend the courtesy of trying the goods yourself first, preferably forty-eight hours in advance.

CHAPTER 7

Perennial Favorites

The beginnings of agriculture around 8,000 BC are generally associated with the big annual cereals, such as the wheat and rice that continue to dominate, but there's some evidence that the first domesticated plant was, in fact, a tree: the fig. Eleven-thousand-year-old fossil specimens located at a house near ancient Jericho reveal seedless (parthenocarpic) fruit. Since seedless figs cannot reproduce on their own, this suggests that Neolithic people cloned them by planting branches, much as is done with figs today. Olives and probably bananas followed around 4,500 BC. Dates, grapes, pomegranates, and sycamore figs were all being cultivated around the Mediterranean by the time of the pyramids, as were cherries in Europe, and peaches and apricots in China. Apples have been harvested—if not exactly cultivated—since the Stone Age. Grafting accelerated the development of the modern apple in ancient times, with the dwarf apple stock reportedly introduced to Europe by none other than Alexander the Great (who wisely chose that moniker over Alex Applestock). And, of course, Adam and Eve weren't kicked out of Eden for munching on chickpeas.

From an agricultural perspective, woody perennials like these have undeniable appeal. Plant or graft, maintain as needed, and then harvest year after year after year. In fact, it might have been

the settling of people correlated with grain cultivation that gave them the dietary stability to tinker with improving woody perennials, as well as the permanent homes near which to do so. Woody perennials have no fewer reasons to recommend themselves to today's urban farmers. Although they are more subject to space restrictions than most herbaceous plants, they hold tremendous potential for integration into city planning. When he was a boy, urban-agriculture evangelist Jac Smit used to tap the maples along the street and in the backyards of his town, boiling down sap to syrup for a little extra scratch. Imagine that on a megacity scale. New York City alone, for example, has over 5,000,000 trees, including nearly 600,000 street trees. If just 5 percent of those street trees were useful ones, such as apple, pecan, or sugar maple, that would be a whole lot of apple brown betty.

Even if you don't own a wooded courtyard or sit in the city planning office, there's a good chance that you can find a woody perennial to fit your farmstead. You will have to do some planning, however, since they tend to be more expensive than annual crops and require a longer time commitment. Your climate zone (see the Zone Map at the back of the book) and choice of cultivar are more important, too, since you are dealing with more or less permanent residents that will need to overwinter at your place. Unlike annual crops, though, a well-planned investment in woody perennials offers years of reward for your toils.

Popular Pomes: Apples and Pears

Unlike a stone fruit with a central pit, such as a peach, apples and pears have many-seeded cores and are known botanically as *pomes*.

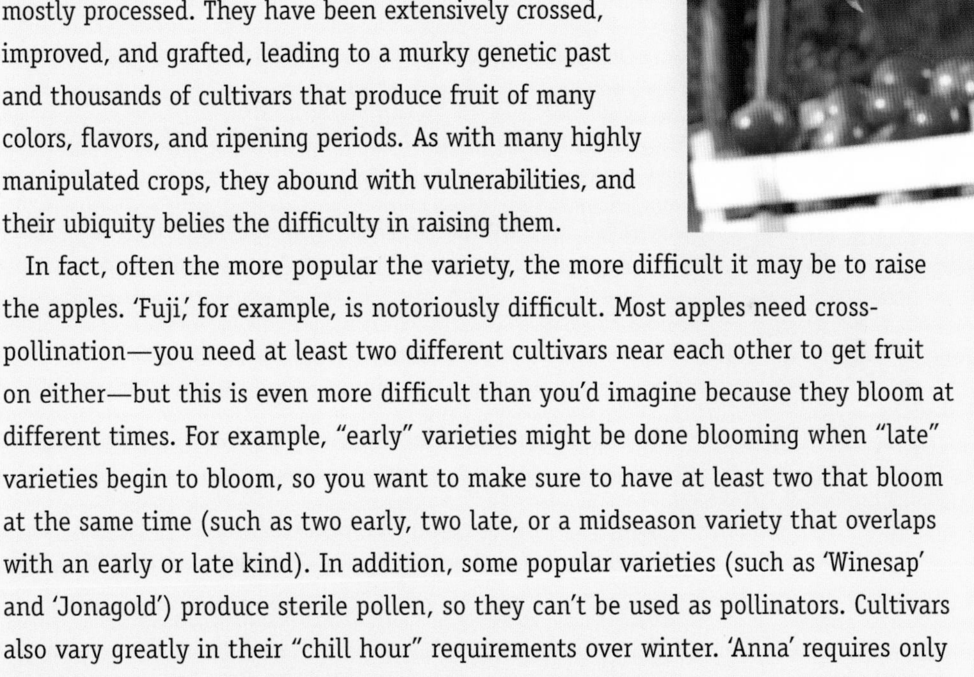

Apples

Whether you're looking merely to keep the doctor away or to expel all of humanity from paradise, apples do the trick. They are perennial favorites among fruit, and Americans each eat nearly 150 of them every year, mostly processed. They have been extensively crossed, improved, and grafted, leading to a murky genetic past and thousands of cultivars that produce fruit of many colors, flavors, and ripening periods. As with many highly manipulated crops, they abound with vulnerabilities, and their ubiquity belies the difficulty in raising them.

In fact, often the more popular the variety, the more difficult it may be to raise the apples. 'Fuji,' for example, is notoriously difficult. Most apples need cross-pollination—you need at least two different cultivars near each other to get fruit on either—but this is even more difficult than you'd imagine because they bloom at different times. For example, "early" varieties might be done blooming when "late" varieties begin to bloom, so you want to make sure to have at least two that bloom at the same time (such as two early, two late, or a midseason variety that overlaps with an early or late kind). In addition, some popular varieties (such as 'Winesap' and 'Jonagold') produce sterile pollen, so they can't be used as pollinators. Cultivars also vary greatly in their "chill hour" requirements over winter. 'Anna' requires only 200 hours or so at temperatures of 45 degrees Fahrenheit (7.25 degrees Celsius) or below, while 'Northern Spy' needs a solid 1,000 hours. Many cultivars will not even bear fruit for several years. Ten or more applications of pesticide per season are not unusual for commercial growers.

Still, clearly apples *can* be grown. If the ancient Romans could do it, so can you. Apples vary greatly in terms of climatic adaptation, but they generally prefer moist, deep, slightly acidic soil. The key is that it drains well. As a class, apples can grow in zones 3–9, although you should work with your cooperative extension to get one suited to your particular microclimate. It would be tragic to plant 'Northern Spy' in Tuscaloosa, for example, and have it live but never fruit successfully. If you're among the lucky few urbanites with enough room to plant a small orchard, more power to you. Check with your cooperative extension for a home orchard publication. If you're more of a dabbler or lack much room, there are a few tricks to help you. They're not substitutes for checking with your cooperative extension, but at least they give you somewhere to start.

First, you'll probably want a dwarf apple tree for reasons of space. Dwarves also mature to fruit production earlier and lend themselves to easy picking (no ladder required). You can find many varieties of apples dwarfed through grafting to one or more dwarfing rootstocks—the same method Alexander the Great apparently brought to Europe. There are a few main varieties, including 'M9' and 'M26.' Dwarf trees grow to be about 8 to 12 feet (2.5 to more than 3.5 m) tall, while semidwarf trees will be about 12 to 20 feet, and standard (undwarfed) trees grow to about 20 to 30 feet (6 to 9 m). I can't give you blanket cultivar recommendations because they perform so differently in different regions, but it's likely that at least one of several highly disease-resistant cultivars—such as 'Liberty,' 'Freedom,' 'Redfree,' and 'Williams' Pride'—will work in your area. All are midseason cultivars, so they can be grown together for cross-pollination. If you buy dwarfed specimens, you could grow them in the ground as trees or prune them to a flat shape ("espalier" them) against a wall or on a trellis (see Chapter 8).

When you buy an apple tree, you're usually buying two different cultivars in one: the scion (for example, 'Jonagold')—the main-event cultivar whose fruit and other characteristics you want—and the rootstock (for example, 'M27'), which is grafted to the scion to improve disease resistance, to produce a dwarfing effect, or to realize other characteristics. You may have a choice of rootstock, but the nursery will often just pick one itself. 'M27' is a very dwarf rootstock that will keep trees to about

6 feet tall or shorter. Growing a cultivar grafted to an 'M27' rootstock is probably best for container culture; just make sure that the central stem has support. There are also a few varieties of columnar apples, which grow 6 to 10 feet (about 2 to 3 m) high but no wider than about 2 feet (61 cm). These can be grown in the ground (to use a space otherwise too small for a traditional apple tree) or in containers (where they will reach about 6 feet tall). As with 'M27'-grafted cultivars, they'll need support. Three commercially available columnar apples are 'Northpole' (midseason, compared to 'McIntosh'), 'Golden Sentinel' (midseason, compared to 'Golden Delicious'), and 'Scarlet Sentinel' (early/midseason, compared to 'Winesap'). In addition to their small footprint, they have the added bonus of requiring little active pruning, unlike most apple trees.

Figure on obtaining at least three different varieties with overlapping bloom times to ensure cross-pollination. Even "self-fruitful" apples will produce better if they have a local cross-pollinator. You don't need to get all three with the exact same bloom time. You could get a mid-season and a late-season, for example, if you also have a mid- to late-season third specimen that overlaps the bloom times of the other two. I've included mainly midseason kinds just to make it easy. Also, the smaller the trees, the nearer they should be to each other for cross-pollination (unless you want to try to cross-pollinate manually). Keep dwarf and columnar apples within 20 feet (6 m) of each other, semi-dwarves within 50 feet (a little more than 15 m), and standards within 100 feet (30.5 m). Crab apples can be successful cross-pollinators of apples, so if you have one growing near your farmstead, you might want to position your apple trees on the (sunny) side of the property closest to the crab apple.

If an early infection develops, you'll need to apply sprays approximately weekly throughout the season.

Companions, Relatives, and Pests

The apple is not-too-distantly related to other popular, large-fruited trees such as the pear, stone fruits, and quince, as well as to ornamental shrubs such as kerria, serviceberries (*Amelanchier* spp.), cotoneaster, and hawthorn—all members of the rose family. Codling moths are the big pest of apples nationwide, joined in the East by plum curculio beetles. Secondary pests include apple maggots across the Northeast to northern Midwest and oriental fruit moths in the South. Conventional sprays depend on where you live, what the local pests are, and when they emerge.

If you're farming on a small scale, which is likely, ecologically friendly approaches are possible. First off, be sure to clear away any dead or diseased vegetation and remove it from your premises. For light codling-moth pressure, pheromone and/or blacklight traps could be the way to go. You may be able to fight plum curculios using chickens, which love to scratch up the larva. If you shake the tree, the adults will fall down and play dead and also get eaten. You can try a spray based on kaolin, which is a fine clay. The spray will leave a chalky film around the fruit and leaves that is effective (although not foolproof) against all four big pests. Another option is to put a plastic or paper bag (or even a piece of pantyhose) around each apple about six weeks after bloom. It seems to have a mixed record, with some people swearing by it and others *at* it.

Disease-wise, the big problem for apple growers is apple scab, a fungus that overwinters on old leaves. The easiest and best solution is to buy scab-resistant apples, such as 'Liberty,' 'Freedom,' 'Redfree,' and 'Williams' Pride.' Fall cleanup of fallen leaves helps for obvious reasons. Even organic production frequently requires sprays, especially in damp weather. Organic sprays typically involve sulfur, lime-sulfur, Bourdeaux mixture (copper sulfate and lime), or compost teas. These sprays or powders need to be used around every rain early in the season to prevent an initial infection and less frequently thereafter.

Pears

Pears are very much like apples in culture. They are closely enough related that you can graft a scion of one to the rootstock of the other. Despite the close relation, pears are blessedly easier to grow, being less bothered by pests and disease and generally more tolerant of poor drainage and humidity. Unlike apples, they come in two major species: the common pear (*Pyrus communis*) and the Asian pear (*P. serotina*). The former is hardy in zones 4–9, the latter in zones 5–9. Both kinds can be made semi-dwarf (12 to 15 feet high [3.5 to 4.5 m) when grafted to a rootstock mysteriously known as 'OHxF 333'. Ask the nursery about this rootstock because it is likely the one to choose. 'OHxF 333'might make the tree

slightly less cold hardy, but the heat island effect and wind protection in a city might compensate for that.

Some pears are fairly self-fertile, but some need cross-pollination, and all benefit from it, so plant three or more varieties. 'Kieffer' is a hybrid of the two species and is self-fertile but benefits from another 'Kieffer' for pollination. Asian and European pears can pollinate each other, but that is only likely for late-blooming Asian pears and early-blooming European ones because Asian pears blossom earlier in the season. It's also possible that ornamental Callery pears (*P. calleryana*), sometimes grown in urban areas as street trees, could cross-pollinate with your pears. Common pears are usually picked before they are fully ripe, while Asian pears are usually left on the tree until full ripeness.

Companions, Relatives, and Pests

Fire blight plays the role of the pear's big nemesis. Choosing fire blight-resistant varieties offers the first and best defense. The 'OHxF 333' rootstock provides some resistance, as do 'Moonglow' (very resistant) and 'Seckel' varieties of the common pear, the 'Shinko' and 'Chojuro' varieties of Asian pear, and the 'Kieffer' hybrid. The bacterium behind fire blight curls tender new growth, then blackens the branch, and then invades the tree, giving it a burnt look. Contemporary pear growers often fight pear blight by spraying the trees with antibiotics. Less drastic options include the use of a Bordeaux mixture on a growth or a product called BlightBan® A506—sometimes in combination with antibiotics—which spreads an innocuous bacterium that competes with fire blight. Good cultural practices call for using only low-nitrogen fertilizer, including manure and organic sources, on pear trees to limit the tender new growth that is the entry portal for fire blight. You should also disinfect pruning tools between uses.

Pear also claims its own variety of scab, which pales in comparison to apple scab and which you can control in the same manner. The pear psylla poses the major insect threat to pears. You can control the insect by attracting natural enemies (such as predatory flies), by applying insecticidal soap during growth, by using kaolin-based spray, or by using sulfur-based sprays *or* dormant oil (not both).

Both kinds of pear are related to the ornamental Callery pear, beloved for the beautiful ornamental 'Bradford' cultivar commonly planted in cities. Callery pears may be able to pollinate your fruit trees. The real problem, however, is that your trees might pollinate *them*. New cultivars of Callery pear have made the once-nonfruiting 'Bradford' grow fruit and become invasive. If you know that Callery pears grow nearby, you might want to check with your local cooperative extension to see if cross-pollination will be a problem.

Bramblefruit and Gooseberries

Brambles are thorny bushes related to roses that generally produce berries on arching canes. They are of the genus *Rubus*, which includes many different species of raspberry, dewberry, and blackberry as well as hybrids such as tayberry, loganberry, and wineberry. Gooseberries are of the genus *Ribes*, appear similarly thorny but are bushier, and share similar cultural requirements. Cape or Chinese gooseberries are an entirely different breed (see Miscellaneous Small Berries, Vines, and Bush Fruit).

Both brambles and gooseberries prefer rich, well-drained, moist, slightly acidic soil, so the regular addition of compost and/or manure is welcome. They are also shallow-rooted, so mulching is a good idea to maintain soil moisture. All benefit from being cut down and switched with new leaders every few years. They fruit most abundantly in full sun but are fairly tolerant of shade (especially gooseberries). Both are heavy feeders and particularly require nitrogen.

Despite their similarities, *Rubus* spp. and *Ribes* spp. are not closely related. Gooseberries are true berries in botanical terms, while blackberries and raspberries are *aggregate* fruit—almost like tiny little clusters of grapes. All bear fruit prolifically and within a year or two of planting. They are excellent permacultural subjects as hedges (the thorns add some extra exclusivity) or edible bushes, and the fruits are famously delicious in jam. Traditionally, gooseberries have also been fermented into gooseberry "champagne" in the United Kingdom.

Raspberries

Raspberry plants are perennial, but the canes are essentially biennial—they die after two years. There are two main varieties classified by when they fruit. *Everbearing* (or *fall-bearing*) cultivars bear once or twice in the season on first-year canes. This means that you can cut them down to the ground every spring or fall to reduce the chance of disease. *Standard* (or *summer-bearing*) raspberries produce fruit only during the summer on second-year canes. You should tie raspberries—particularly standard cultivars—to trellises or other support 2 or 3 feet (61 to 91 cm) off the ground.

To prune standard raspberries, cut fruiting canes and wimpy first-year canes after harvest and then tie strong first-year canes to the trellis for overwintering. The next spring, thin the newly forming canes and repeat the process. Everbearing raspberries present even less fuss to manage: simply mow them down to the ground in late winter or early spring and thin newly forming canes when they emerge.

Everbearing raspberries probably adapt best to urban farming. They need little (if any) support, are simple to prune, and resist disease better since you eliminate mature canes in the prime of life. 'Heritage' (red, late-bearing) and 'Anne' (yellow) are popular everbearing cultivars adapted to a wide geographic range. If you want to try summer-bearing kinds, 'Meeker' (red, popular in the Northwest) and 'Prelude' (red, very early bearing and disease resistant) are good to check out first.

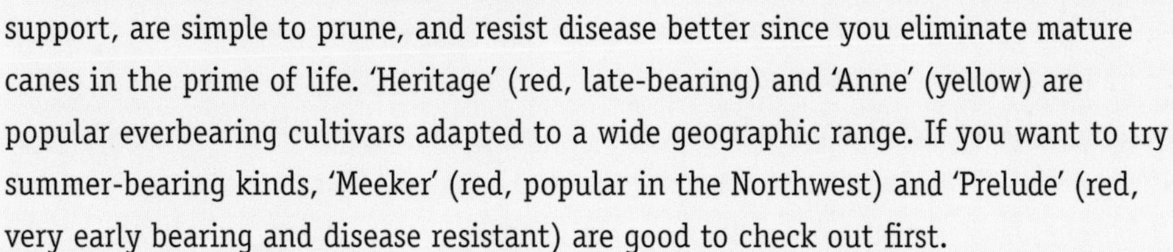

Blackberries

There is actually a cluster of different species and subspecies within the genus *Rubus* known as blackberries (or dewberries), with the most notable being *R. fruticosus*. Even more confusing, unripe blackberries are red, and some raspberries are black. A simple way to distinguish them is to pull off a ripe fruit. Raspberries tend to leave the

receptacle behind, becoming a little cup, while blackberries take it with them, giving them a solid core. Blackberry bushes can be either tall and erect or trailing and long. Having deeper roots than raspberries, blackberries are slightly more resistant to drought. They are usually cut down annually like everbearing raspberries. While trailing kinds can spread for yards and require support, erect cultivars need little, if any, support. 'Darrow' is a general all-around recommendation. 'Brazos' and 'Arapaho' (which is nearly thornless) are particularly suited to the South, and 'Illini' to the far North. All are erect varieties.

Gooseberries

If you've never had a gooseberry, you're missing a treat. The bite-size fruit offers a nice blend of sweet and tart and appears in abundance. The green ones look like miniature watermelons. As with so many fruits, gooseberries come in Old World and New World varieties, the two most important species being European gooseberry (*R. grossularia*) and American gooseberry (*R. hirtellum*). The European ones are generally larger, and the American ones are generally tougher. Favorite cultivars include 'Poorman,' (red American cultivar with large fruit; mildew resistant), 'Early Sulphur' (yellow European cultivar with medium fruit; susceptible to mildew), and 'Hinnonmaki Yellow' and 'Hinnonmaki Red' (hybrids with medium-sized fruit; mildew resistant). Gooseberries (zones 3–5) withstand cold better than the bramblefruit, followed by red raspberries (zones 3–9), and then blackberries (zones 4–9), and then black raspberries (zones 5–9, the most heat tolerant).

Companions, Relatives, and Pests

Raspberry and/or blackberry hybrids such as loganberry and boysenberry are occasionally available, too. They are often thornless and usually grown much like blackberries. Brambles are a promiscuous bunch, hybridizing easily, so it can be difficult to know just what kind you have unless you buy an identified variety. For bramblefruit, avoid, if possible, soil that has held plants susceptible to verticillium wilt and/or *Phytophthora* within the past five years; these include nightshades, strawberries, cucurbits, apples, peaches, and cherries.

Starting near the beginning of the twentieth century, gooseberries were banned and actively eradicated in many parts of the United States for fear of spreading the devastating white pine blister rust. The fungus that causes the disease, introduced to the United States around 1900, has a strange life cycle that requires it to bounce between pines and an alternate host, such as gooseberry. Without both hosts, it cannot survive. When this connection was discovered, gooseberries and related *Ribes* spp. were sacrificed on the altar of the white pine industry. Since then, resistant pines and gooseberries have been developed, and scientists have discovered that alternate hosts must be within about 1,000 feet of pines to pose a threat (gooseberries can carry the rust, but aren't affected by it). Restrictions remain where the pine industry is economically important (or where legislative change is glacial) and have been relaxed in other areas. The most direct way to find out about your area is to ask your local cooperative extension.

Barring any restrictions, you can take a two-part approach. First of all, reputable mail-order nurseries selling gooseberries will say up front that they won't ship to specific states; see if your state (or county) is one of them. Second, search online for "noxious weeds" and "ribes" and "[your state name]" to see if anything shows up. If a mail-order outfit will ship to you and, your Internet search turns up no restrictions (and you're not smack in the middle of a pine forest), you're probably safe.

Easily rooting where their branches touch the ground, raspberries and blackberry varieties become invasive, arching across your property like a sea serpent. Along with gooseberries, they can also spread by seed, so you should keep a close eye on all of them to keep them from becoming weeds. Brambles are related to roses and strawberries, and gooseberries to all kinds of currants (except zante currants, which are actually grapes).

Citrus

Members of the citrus family, Rutaceae, are beloved the world over—orange juice is traded like oil on commodity exchanges, after all—even though most citrus's climatic preferences limit commercial production to tropical and subtropical regions. Despite their limited climatic preference, most citrus plants are otherwise not picky as long as they have well-drained soil and full sun. If you live in the warmest parts of the United States or have your own personal greenhouse, yours is a huge range of citrus options, from Australian finger lime to yuzu, a particularly fragrant citrus fruit favored in some Asian cuisines. Fortunately for those of us in colder areas, citrus are also excellent plants for containers. There are two main options available to most every corner of the United States: inherently compact plants that can be grown in the ground or in containers, or any-size citrus grafted to a dwarfing rootstock, such as *Poncirus trifoliata*.

A citrus blossom

Calamondins and Kumquats

Let's talk about small plants first. Calamondins (*X Citrofortunella mitis*, one of those strange intergeneric hybrids) and kumquats (*Fortunella* spp.) both produce small, intriguingly sour-and-sweet fruit. Calamondins look like tiny oranges and are usually sold as the species (not as cultivars). Kumquats often resemble grape-shaped oranges and come in named varieties synonymous with their species; 'Marumi'

(*F. japonica*) and 'Nagami' (*F. margarita*) are probably good bets. Calamondins and kumquats are among the most cold hardy of edible citrus species, which is handy in the case of unexpected frosts.

Limes and (a) Lemon

Limes are another option, including the Persian, Key, and Kaffir varieties. Persian (or Tahiti; *Citrus × latifolia*) limes are the kind you usually see in supermarkets, and 'Bearss' is a common seedless cultivar. Mexican limes (*C. aurantifolia*) are a more pungent cousin (also known as Key limes of pie fame). Kaffir limes (*C. hystrix*) may be the most reliable of the three, since the variety is typically grown for its leaves—essential to Southeast Asian cuisine—more than for its fruits. None of the three handles cold well, though you can also get various kumquat/lime crosses (called "limequats") with a slightly hardier disposition. There are three main cultivars: 'Eustis' (the most popular), 'Lakeland,' and 'Tavares.'

An overwintering Kaffir lime

For those preferring lemons, 'Meyer Improved' is the commonly recommended variety for containers. Like limequats, it is actually a container-friendly hybrid (*Citrus × meyeri*) that is somewhat more cold tolerant than regular lemon (*C. limon*).

Oranges

Oranges and other orangey citrus are larger plants but not impossible for container culture. For true sweet oranges (*C. sinensis*), the best container cultivars are generally thought to be 'Washington Navel' (seedless, winter ripening) or, better yet, 'Trovita,' a likely variation on 'Washington Navel' that ripens in spring and is also seedless. Mandarin oranges are actually a different species, *C. reticulata*, that

is hardier than sweet oranges. Satsumas are a particularly hardy type of mandarin, and 'Owari' and 'Kimbrough' are hardy even for satsumas and among the best suited to container culture. Both ripen around November and December and both are usually seedless.

Even with these relatively small citrus trees, you will have to move container-grown specimens to larger pots every few years until you hit about 15 or more gallons in volume. The best bet is to buy an orange cultivar such as 'Trovita' grafted to a *Poncirus trifoliata* rootstock, a combination that is not too difficult to find. Terra-cotta is often recommended (but remember: it's heavy), as is plastic, as long as the pot is light in color or shaded enough to keep from overheating the roots.

P. trifoliata, also known as trifoliate orange, is the tough guy of the citrus world. It is exceptionally hardy for citrus (down to at least zone 6) and conveys some of that cold tolerance (and possibly some disease resistance) to citrus grafted onto it. Ungrafted, it grows something like a living wall of barbed wire, adorned with long thorns (curved in the 'Flying Dragon' cultivar) and bitter fruit. It is sometimes grown as an impenetrable hedge, which might serve many urban farmers' purposes well, irrespective of its grafting potential. You can buy other citrus grafted onto trifoliate orange, as well, to dwarf it for container growing and/or to provide some additional cold-hardiness.

Citrus plants prefer fertilizer with roughly twice as much nitrogen as phosphorous and potassium (such as 12–6–6). They are also vulnerable to micronutrient deficiencies, so make sure that your fertilizer contains micronutrients or that you supply them another way. You can easily find citrus-specific fertilizer mixes with good NPK ratios and the essential micronutrients. Seaweed-based foliar sprays are a good idea, particularly for container-grown plants. Citrus like to be watered deeply but infrequently, but never allow them to dry out. Trees in containers will probably need to be drenched at least once a week.

Companions, Relatives, and Pests

Like apples and other long-cultivated popular fruit trees, citrus are rife with invertebrate pests. Your best strike in this long battle is to buy certified disease-free plants from reputable dealers; your second-best defense is to keep the trees as healthy as possible through proper care.

Three of the big citrus pests are nematodes, aphids, and scale insects. You can't do much about nematodes, which is one reason to buy from a reputable dealer and avoid importing them. Growing citrus in containers using sterilized potting medium also minimizes the risk of nematodes. Aphids look like tiny water balloons with legs and are usually found feeding on the undersides of leaves, causing them to curl. They can also, although rarely, transmit a citrus virus. Scale insects look like unmoving little shells stippling the tree. Both aphids and scale insects produce a sticky substance called *honeydew* (unrelated to the melon), which provides a clue to their presence. Honeydew can promote sooty mold, which you don't want, and attract ants, which you also don't want. If you see ants walking up and down a tree, in fact, they are probably farming aphids or scales. The ants use the honeydew as food and, in exchange, protect the pests from predators and parasites. This is called *trophobiosis* in scientific terms and a "protection racket" in layman's terms.

Getting rid of the ants is critical. Setting out ant-bait traps is one step, and circling the trunk with a sticky substance (such as Tanglefoot®) is another. The University of California Statewide Integrated Pest Management Program also mentions trying a slippery barrier like Teflon™, but my ants seem to have no problem crawling over it. Once the ants are eliminated or kept at bay by a barrier, the scale insects and aphids are defenseless to predators. You can blast aphids off using a water spray, and you can remove scale insects using cotton swabs dipped in rubbing alcohol. Try to then maintain control by regularly spraying the plants with a mixture of mild dish detergent and water (about 2 tablespoons [30 ml] detergent to 1 gallon [about 4 liters] water).

Diseases pose less of a threat to citrus. They may reduce yield or impair attractiveness, but they seldom kill the plant (although commercial growers will certainly burn the plant if there's a risk of spreading disease). Sooty mold is mainly disfiguring. You can remove it by gently scrubbing the leaves with a disposable cloth dipped in the detergent solution, but you can eradicate it only by eliminating the honeydew source.

Good Neighbors in Charlotte

The garden at the interfaith Urban Ministry Center (UMC) in Charlotte, North Carolina, is a testament to second chances, from its repurposing of the historic Seaboard Coast Line Railroad Company Office Building to the countless small dignities it provides the homeless "neighbors" it serves so that they can enjoy a new and fuller life. For thousands of these neighbors, UMC is about more than a reliable meal, a hot shower, or referrals to services. It is about fellowship—nowhere more so than at UMC's community garden.

It's not what you'd call a formal garden. There are wall-size murals, a labyrinth of painted rubble, a bottle tree (imagine a coat rack with empties), and other works of art, such as two orange-painted bicycles. And, of course, there's an odd mixture of plants, among them patches of tomatoes, fig trees, a single stalk of corn, burdock and castor oil plants, an unusually formal grouping of blueberries, beautiful watermelons, and lots and lots of sunflowers. The garden may only provide a small fraction of the 100,000 meals UMC serves its neighbors each year, but its spiritual worth far outweighs its caloric value. Seldom have I seen a garden with so much meaning.

With a transient population of urban farmers and more than a handful of the Center's 10,000 volunteers, it is—as Garden Director Don Boekelheide puts it—"a great teacher in letting go." There was, for example, the volunteer who unknowingly mowed down the wildflower meadow—before giving it a spray of herbicide for good measure. And the guy who trapped a tomato-munching squirrel in his home garden, drove it over to UMC, and compassionately released it into the community garden, where it made a beeline for their tomatoes. And there's always what Boekelheide calls the "Darwinian" lawn, surviving without much care and always looking for an untended spot to exploit among the ornamentals and food crops.

Don Boekelheide (center), flanked by sometimes farmers Harry and Cleo, next to the bottle tree.

"We tolerate weediness," says Boekelheide, "in exchange for soil health and overall environmental health." It's as good a metaphor as any for a place that welcomes people who are unwanted most everywhere else. There's even larger meaning in the choice to grow food organically, according to Boekelheide. "We discourage people from becoming chemically dependent, and our garden simply follows the same philosophy."

Miscellaneous Small Berries, Vines, and Bush Fruit

Many of us regard a berry the same way former Supreme Court Justice Potter Stewart regarded hard-core pornography: "I know it when I see it." Speaking botanically, however, there's really a huge variety of specific fruit types depending upon the very private relation of a plant's edible parts to its ovaries. As a result, many things not commonly considered berries (such as tomatoes and grapes) are berries, and many things spoken of as berries are technically something else, such as aggregate fruit (blackberries, raspberries), drupes (hackberries), epigynous fruit (blueberry, lingonberry, cranberry), accessory fruit (strawberry), and multiple fruit (mulberry). Fortunately for the urban farmer, some of the hardiest woody perennials produce delicious berries (or whatever they're called) and other wonderful small fruit.

Strawberries

Like bramblefruit, strawberries are actually a group of closely related species of a single genus of the rose family (in this case, *Fragaria*). The modern strawberry is also one of North America's proudest contributions to the world diet in that it is a cross of our native *F. virginiana* with *F. chiloensis*, a strawberry from South America, which claims so many additions to the world diet (tomatoes, peppers, chocolate) that it doesn't need to brag about strawberries. Woodland or alpine strawberries (*F. vesca*) occur throughout the Northern Hemisphere.

People have long tinkered with the modern strawberry (known as *F.* × *ananassa*), producing varieties that are June-bearing, everbearing, day-neutral, short-day, and long-day, the first three of which concern us most. June-bearing strawberries bear their entire crop, which is relatively large, at a particular time of year (usually around June). This contrasts with everbearing strawberries,

which are somewhat misnamed. Rather than continuously producing, everbearing strawberries tend to produce an initial spring crop and a smaller fall crop.

Day-neutral strawberries are how you'd expect "everbearing" strawberries to be, producing continuously—about every six weeks—as long as it's not too hot or cold. A related strawberry worthy of attention is the alpine strawberry (*F. vesca*), another species that produces small berries with big flavor. It is often treated as an ornamental plant (edging fixtures or ambling along a garden path) that happens to produce occasional treats, rather than as a real production berry.

Strawberries are an excellent option for an early-season fruit crop, as a living mulch for trees and leggy bushes, and for intercropping. They have long been favorites in vertical-intensive applications, such as strawberry barrels, strawberry jars, and hanging baskets.

Well-drained, slightly acidic soil rich in organic matter is best, but strawberries are more forgiving of soil conditions than blueberries and some other fruit are. Strawberries should be watered regularly and mulched because they have very shallow roots. All varieties should be given space to spread, although this is less vital with alpine strawberries, which are very compact and can be spaced more closely together. Alpine strawberries also prefer partial shade, whereas the other kinds need full sun. All varieties of strawberry tolerate a wide range of climate zones, generally at least zones 4–10 for June-bearing and alpine strawberries and zones 4–8 for everbearing and day-neutral varieties.

Companions, Relatives, and Pests

By themselves, strawberries are actually quite a wimpy bunch, easily pushed aside by weeds and other pests. They are troubled by a number of diseases and fungal conditions, so buy resistant varieties that are virus indexed (which means that they are progapaged from virus-free stock) and/or certified to be disease free. Commercial production usually limits any given plant to three years of production to ensure vigor, and treating strawberries as annuals is common in warmer regions. In fact, strawberry growers often practice plasticulture, growing strawberries as annuals through vast sheets of plastic that control weeds and pests. It is common in Florida for growers to interplant strawberries with cucurbits (cucumbers, squash, and melons) near the end of the season to take advantage of the existing cultural conditions. Alliums such as leeks, onions, and chives are often recommended as companion plants, as alliums are a hardier group, do not compete much for light or soil nutrients, and may even repel some strawberry pests.

In Pakistan and elsewhere, strawberries are often planted around orchard trees that have not yet reached fruiting age. Peach orchards in New Jersey frequently interplant with strawberries not just for efficiency but also because a parasitic wasp that attacks oriental fruit moths—a pest of peaches and other fruit—likes to lay its eggs in the hibernating larvae of strawberry leaf rollers, a major pest of strawberries. It's a two-fer. This is a good example of integrated pest management (see chapter 8).

Grapes

It would be hard to tell the human story without reference to grapes. These wondrous vine fruits can be trained against a wall, they bear heavily, and they provide the grower with one of the best opportunities to make wine (or vinegar, depending on skill level and interests). There are many varieties of grapes with different habits, tastes, and qualities, including the 'Concord' grape of peanut-butter-and-jelly fame (*Vitis labrusca*), the giant southern muscadine grape (*V. rotundifolia*), the European wine grape (*V. vinifera*), and dozens of other native and exotic species. We'll focus on the first three.

Grapes are, perhaps, the best known of the vine fruits. The fruits are eaten fresh, boiled into jams and jellies, dried into raisins, and, of course, fermented into wine. Grape wood is gnarly and used for many ornamental purposes and for smoking food. Many cuisines employ grape leaves to wrap foods, such as in Greek dolmades. Deep rooted and long lived, they are renowned for their ability to live in hardscrabble soils, from the chalky hills of Champagne to the arid expanses of South Africa's Namaqualand. In fact, it is these geological variations, along with microclimates, that comprise the terroir reflected in good wines. Grapes are excellent for growing along fences and

A 'Chardonnay' grape plant being cultivated along a balcony rail.

walls or overhead on pergolas and other structures.

Grapes favor deep, well-drained soil, and they fruit best in full sun. American grape varieties (zones 4–10 for *V. labrusca* and 6–9 for *V. rotundifolia*) are hardier than European grapes (zones 6–10), although there are European hybrids that are hardy down to zone 5. The vines will go all over the place if allowed but are generally trellised for commercial and home production. You should buy cuttings or vines from named cultivars for each grape species. Popular varieties include: 'Concord' (*V. labrusca*, zones 4–8, a popular purple grape for juice and grape jelly), 'Carlos' (*V. rotundifolia*, zones 6–9, a self-fertile bronze muscadine), 'Merlot' (*V. vinifera*, zones 6–10, a premier grape for red wine), and 'Chardonel' (European hybrid, zones 5–8, a wine cultivar similar to but hardier than 'Chardonnay').

Companions, Relatives, and Pests

The archenemy of European grapes (*V. vinifera*) is *Phylloxera*, which is sort of like the root aphid. European vines are therefore usually grafted onto rootstocks of American species, which are resistant. As long as you grow certified disease-free American cultivars or grafted-hybrid European cultivars and allow for good air circulation, the major pests are usually just birds and wasps. If other invertebrates or fungal diseases (such as downy mildew) seem to be posing a problem, check with your local cooperative extension on how to target those particular pests.

There are a number of wild grapes that look suitably grapelike. Grapes are also related to Virginia creeper (*Parthenocissus quinquefolia*). Commercial vineyards often grow roses at the end of grape rows, ostensibly as a disease sentinel or to attract bees, but the utility of the practice is unproven.

Figs

Perhaps no fruit has quite as cultish a following as the common fig (*Ficus carica*). It is quite possibly the first domesticated plant, with its mysterious fruit enjoyed since antiquity, and its distinctive leaves the chosen underwear of Adam and Eve.

In the United States, fig production is concentrated in California's Mediterranean-like climate, and world production around the Mediterranean Sea itself. Although figs grow best in zones 8–10, some will survive down to 5 or 6, and there are obsessive fig fanciers everywhere in the country who wrap the trees with burlap or bury them every season to survive the winter. I'm in zone 6 and am experimenting with a half dozen varieties.

There are hundreds of cultivars of fig, each suited to its own purpose, whether for eating fresh, drying, or preserving. For most urban farmers, however, there are probably three main criteria: self-fertility, cold-hardiness, and size. Self-fertile (*parthenocarpic*) cultivars are important in the vast majority of the country because of the absence of fig wasps, the trees' natural pollinator. There are plenty of self-fertile choices, however. Cold-hardiness is obviously important to northern gardeners, but many cold-hardy varieties can also grow well (or better) in more southern regions. Finally, some cultivars are naturally smaller than others and thus better suited to container culture or cramped urban spaces. Four varieties that match all of these criteria are 'Hardy Chicago' (dark, good in the North), 'Celeste' (dark, good in most places but the Southwest), 'Conadria' (light, good in the Southeast and California), and 'Violette de Bordeaux' (dark and small, good throughout the West). You should check with your cooperative extension, however, since some will give you state-specific recommendations. Figs prefer full sun and well-drained soil, and they may produce two crops: a small one (called the *breba crop*) followed by the larger, main harvest.

Companions, Relatives, and Pests

Birds and frost are probably the biggest enemies of figs, which are distantly related to mulberries. Netting the trees during fruiting will help deter birds, while fig trees can be protected from frost in colder areas by moving potted specimens to unheated, sheltered sites; wrapping them in burlap; or digging each tree halfway out, laying it down, and mounding it with soil for winter. Even trees apparently killed from cold can often regrow from roots, although it's not safe to assume this. You can treat most other insect or pathogen-related problems—such as scale insects, ants, or nutrient deficiencies—as you would for other trees.

Kiwifruit

Fruiting vines of the genus *Actinidia* bear one of those fruits that just can't escape misnomers. They're often called *kiwifruit* for their popularity in New Zealand (where they are not native) and sometimes called Chinese gooseberries (they are native to temperate forests in East Asia, but they're not gooseberries).

Unimproved kiwi vines are *dioecious* (there are male plants and female plants), but some modern kinds will bear fruit without partners. There are three main cultivated species, each with commercially popular cultivars. All are somewhat shade tolerant and need very strong support (which trees provide where they grow wild). They like moist, well-drained, slightly acidic-to-neutral soil, and they are heavy nitrogen feeders. Kiwi vines can also be grown in large containers.

Kiwi buds are susceptible to late spring freezes, so it's probably a good idea to avoid a southern exposure. They also may not pollinate reliably. Female flowers produce no nectar, so bees just aren't that into them. Commercial growers usually rely on saturation pollination, by which hives are put in the center of a kiwi crop, leaving the bees little choice but to go to the kiwi blossoms. You might just try helping out with some hand pollination using a brush. In addition to fruit and pretty foliage, twisty kiwi vines are sometimes used as interesting specimens in the florist trade.

A. deliciosa (*syn. chinensis*) is the fuzzy supermarket species. 'Hayward' is the standard cultivar, but it requires a male companion; 'Blake' is a self-fertile variety. Both are hardy in zones 7–9 and can reportedly grow 25 or more feet (more than 7.5 m) in a single season.

A. arguta (known as hardy or Siberian kiwi or Tara vine) is hardy in zones 4– 8, producing 1-inch, smooth-skinned fruit. 'Issai' is self-fertile, and 'Ananasnaja' is a female, bearing fruit generally described as sweet and spicy. 'Ananasnaja' requires an anonymous male *A. arguta* pollen donor, which will also increase yields with 'Issai.'

A. kolomikta (kolomikta vine) is the smallest and hardiest species, growing in zones 3–8 and with pretty, variegated leaves and grape-sized fruit. 'Arctic Beauty' is the most common recommendation, though you need at least one male 'Arctic Beauty' for the female 'Arctic Beauty' to bear fruit. One male will suffice for every eight or so females. With its small size, extreme hardiness, and pretty foliage, *A. kolomikta* holds a lot of potential for urban growers.

Companions, Relatives, and Pests

Other than late spring frosts, all cultivated kiwi species are tough vines, generally untroubled by diseases or pests—with one proviso. Some kiwi vines, particularly *A. kolomikta*, smell reminiscent of catnip and attract cats. If you dislike cats, consider them as another possible pest (although more to you than to the plant). Then again, if you like cats or have a separate rodent problem, kiwi vines just might be perfect. There are many related *Actinidia* species, and the genus is distantly related to the blueberries, cranberries, and rhododendrons, in addition to other members of the order Ericales.

Blueberries

Blueberry may be one of the perfect plants for urban farms in North America. It bears fruit year after year, tolerates some shade, and looks great, and many varieties are native—the term *blueberry* is used for many different species in the genus *Vaccinium*. Blueberries regularly make top-ten lists of the healthiest foods. If all the hype (and a lot of research) is to be believed, blueberries can protect the heart, brain, and urinary tract as well as help prevent cancer. They are good sources of fiber, manganese, and vitamins B6, C, and K, but the real reason for their good reputation is that they are extraordinarily potent providers of antioxidants and other phytonutrients. Blueberries are freighted with them, from anthocyanins to pterostilbene, making the little blue spheres about as close to wonder-pills as we know.

Blueberries—especially dwarf varieties—are well-suited to containers because they have an unusually extreme cultural requirement: acidic soil, in the range of 4.0–5.5, or about somewhere in between the acidity of orange juice and coffee. That's far more acidic than you'd find in a commercial bag of soil, so you'd need to amend any mix with a lot of organic matter (particularly acidic additives such as peat moss, softwood sawdust, or pine needles) and sulfur to bring it down to the right pH. Blueberries also need special fertilizer (conventionally, ammonium sulfate) to keep the pH down. They prefer well-drained, moist soil and benefit from good-quality mulch. Blueberries can produce a small crop (a couple of pints) in the third year after planting and, with proper pruning, will produce successively more over the following few years until they reach their peak.

There are four main varieties of blueberry: Northern highbush (*Vaccinium corymbosum*, hardy in zones 3–7), rabbiteye (*V. ashei*, hardy in zones 7–9), Southern highbush (*V. corymbosum* × *V. darrowi*, hardy in zones 7–10), and wild or lowbush (*V. angustifolium*, zones 2–5). Although some cultivars are self-fertile, you should plant two or more cultivars of any one variety (Northern highbush, rabbiteye, Southern highbush, or wild) to ensure pollination. A few self-fertile dwarf cultivars well suited for containers include 'Northblue' (a wild/Northern highbush hybrid), 'Tophat,' and 'Northsky' (both *V. angustifolium*). As with full-size plants, companion cultivars will provide better yields.

Two 'Bluejay' Northern highbush blueberries grow in terra-cotta at the People's Garden at USDA headquarters.

Companions, Relatives, and Pests

People and wildlife alike love blueberries—which means covering them with a net if you want as many pints as possible, or leaving them uncovered if you want to share with your phylogenetic cousins. Blueberries also face other pests, although they are not as troubled as longer-cultivated fruits, such as apples. If you do find something ailing your plants, however, check with your local cooperative extension regarding treatment of any locally prevalent pests.

Blueberries are related to other *Vaccinium* species; all members of the heath family (Ericaceae), including lingonberry (*V. vitisidaea*), cranberry (multiple *Vaccinium* spp.), and bilberry (multiple *Vaccinium* spp.); and American wintergreen (*Gaultheria procumbens*). All have similar cultural requirements and could be paired with blueberries. Wintergreen would make a good match with highbush or hybrid blueberries; it could help serve as a living mulch. Wintergreen also tolerates full shade, making it ideal for underplanting—and, of course, it smells great! Blueberries are also distantly related to kiwi vines and other members of the order Ericales, such as rhododendrons and azaleas.

Ground Cherries

Ground cherries (*Physalis* spp.) are also known as cape gooseberries (usually *P. peruviana*, about 3 feet tall) or Cossack pineapples (usually *P. pruinosa*, about half that), the former for their appearance and the latter for their taste. Like tomatillos (*P. philadelphica, syn. ixocarpa*), they produce small, round fruit in papery husks that resemble lanterns. They can be dried into a form akin to raisins. Ground cherries share the same basic cultural requirements as tomatoes (see "The Usual Nightshades in chapter 6) and are excellent candidates for container culture. An inedible relative, *P. alkekengi*, is the ornamental plant known as Chinese lanterns.

Companions, Relatives, and Pests

Ground cherries are closely related to tomatillos, tomatoes, and other members of the family Solanaceae, which also includes potatoes, eggplant, and tobacco. Some species of Physalis can become invasive, especially (*P. alkekengi*), so check with your state's noxious weeds list and keep an eye on your ground cherries, which can self-seed.

Prickly Pear

Prickly pear is not truly a pear, but it sure is prickly. It is actually the fruit of the native American cactus genus *Opuntia*. Opuntias are classical paddle-looking cactuses, and they're surprisingly hardy. Eastern prickly pear (*O. humifusa*), for example, naturally grows from New Mexico across to Florida and all the way up to Ontario. The main opuntia in agriculture, however, is *O. ficus-indica*, "the" prickly pear or Indian fig. It has a far narrower range than *O. humifusa*—pretty much along the southern coastal states from California in the West and through North Carolina in the East—but it can be grown in large containers, too.

In addition to being a tough and ornamental plant, both the fruits and the young stems (the pads, called *nopales*) are edible. Is it the perfect crop? No. The fruits are sweet but seedy, and the whole plant is covered with spines. Worse than the big, fixed spines are the easily detached tiny ones called glochids. They are like little harpoons, and they must be burned off or abraded before prickly pears are eaten.

If you know what you're dealing with, however, you have a perennial, low-maintenance crop that will provide you fruits and green vegetables for years to come. It is salt and drought tolerant and can be grown into a hedge in the southern United States. There are even spineless cultivars, although purists scoff at them. If well cared for and fed high-nitrogen fertilizer within zones 8–10, each plant reportedly can provide 50 to more than 100 pounds (23 to more than 45 kg) of nopales annually. That's a lot of green vegetable, not to mention the fruit.

Companions, Relatives, and Pests

Opuntia spp. are relatively pest-free. When you look at all of the ills facing more popular fruit, spines don't seem so bad. *Opuntia* are closely related to chollas and more distantly to other cactuses, including dragonfruit (pitaya, *Hylocereus* spp.).

Some Exotic Natives

Apples and pears are perennial favorites, but another tack to take either for oneself or for marketing would be to grow something a little less conventional—preferably something with a short enough shelf life to make it competitive with imported produce. Good possibilities include pawpaw, red mulberry, and persimmon.

Pawpaw

Pawpaw (*Asimina triloba*) seems awfully exotic, but is entirely native. In fact, it's the largest fruit native to North America. Pawpaws look like small mangoes or stubby bananas, and they are favorably compared in taste to bananas or other tropical fruit, custardy in texture. (Like the tropical durian fruit, however, they can simultaneously smell awful.) One reason they are not commonly eaten—besides the fact that early colonists preferred familiar fruits, such as apples—is that they neither keep nor travel well. What is bad for commercial production, however, may be good for a local urban grower who can develop a market for the fruit.

A size comparison of a pawpaw with an average-sized apple.

You need at least two different cultivars or unnamed seedlings for pollination, and it may take five years before you see fruit. But that's just a five-year head start on the competition, right? Pawpaws are pretty, small trees 30 feet (9 m) or less that are tolerant of some shade, particularly when young. They are sold either as unnamed seedlings or as one of a relatively small number of cultivars.

Pawpaws thrive in zones 5–8 with 32 inches (81 cm) or more of rain each year and are natively concentrated in a triangle extending from Ontario down through Texas to the west and Florida to the east, excluding Minnesota and New England. They don't even need bees for pollination—they use flies. The perfect urban use within this pawpaw triangle would be as a collection of (relatively) pest-free, disease-free, small shade trees. That way, even if you don't get much fruit, you have shade. And if you do get fruit, you get to taste a forgotten treat and most likely be the only pawpaw supplier in the city. Pawpaws prefer moist, rich, well-drained soil with a pH of about 6.4 but tolerate less ideal conditions.

Companions, Relatives, and Pests

Pawpaws are basically untroubled by diseases or pests. One of its few pests are Zebra swallowtail butterfly caterpillars, which eat *Asimina* spp. exclusively. It's unlikely you'd attract enough to ravage your tree, however, and if you're going to have a pest, it may as well be a gorgeous butterfly, right? *A. tribola* is related to several other *Asimina* species native to North America and is an uncharacteristically cold-hardy member of the custard apple family (Annonaceae), which contains many important tropical fruits, such as cherimoya and soursop.

Red Mulberry

Red mulberry (*Morus rubra*) has a native range overlapping and just slightly larger than that of pawpaw. It is not as popular as white mulberry (*M. alba*), an East Asia native that is the main food of silkworms. Both are tough trees, freely hybridize, and survive urban conditions in zones 4–8. White mulberry was originally popularized in hopes of starting local silk industries and subsequently adopted as an ornamental tree. Red mulberry has delicious fruit resembling an elongated blackberry and should probably be preferred to the white mulberry—which is so tough and prolific that it has become a tree weed in many places—and the also delicious black mulberry (*M. nigra*), which is neither native nor quite as hardy.

Mulberries like well-drained, moist, deep soil—not necessarily very rich—but can tolerate some drought once established. There are a few cultivars of *M. rubra*, including 'Johnson' and 'Travis,' available, and many more of other *Morus* spp. and their hybrids. The red mulberry is a medium-sized tree, growing up to 70 feet (almost 21.5 m) or so under ideal conditions but probably much shorter in a city. The California Rare Fruit Growers also lists it as being good for culture in large containers. Mulberries begin bearing fruit after four or more years. Be aware: birds love them.

Companions, Relatives, and Pests

Other than birds, few pests bother mulberries (although silkworms like them a lot). In addition to each other, red, white, and black mulberries are related to dozens of other *Morus* species and are parents of countless hybrids. They are distantly related to figs, jackfruit, and breadfruit.

Persimmons

Persimmons are another temperate tree fruit with both Asian and North American members. The Asian persimmon (*Diospyros kaki*, sometimes called Japanese or Oriental) is the main commercial variety and involves thousands of named cultivars. It has large fruit that keeps well, and it grows in zones 7–10 in deep, moist, well-drained soil of nearly neutral pH. The American persimmon (*D. virginiana*) is hardier (zones 4 or 5 through 9) and often provides a rootstock that makes Asian varieties more cold-hardy. It has smaller fruit that does not keep well but which is often described as even more flavorful. Both the Asian and American varieties are notoriously astringent when unripe, although some nonastringent Asian varieties have been developed.

American persimmon trees can reach upward of 50 feet (15 m) in the right conditions, but they are slow growers. Asian persimmon trees are somewhat smaller. Some Asian persimmons are self-fruitful, but all American persimmons (and most Asian ones) are dioecious, with both male and female members. Only female trees fruit, but both are needed for pollination. 'Jiro' and its variants are popular, self-fertile, nonastringent Asian cultivars, firm and eaten like apples (also cold hardy to zone 6). 'Hachiya' is a self-fertile astringent cultivar that ripens to a sweet, pudding-like mush. There are many other varieties to choose from. American persimmon has fewer named cultivars, but 'Meader' is a reportedly delicious, self-fertile cultivar with a maximum height of about 25 feet (7.5 m). 'Szukis' is an excellent pollinator that occasionally bears fruit itself. Both are very cold hardy, even for American persimmons.

Companions, Relatives, and Pests

Some persimmons, such as this 'Hachiya,' cultivar, need to be painfully, bruisingly ripe before they are edible. This one isn't ready yet.

Persimmons have few invertebrate pests other than the occasional unholy alliance between ants and mealybugs or scale insects, which can be controlled by circling the trunk with sticky products such as Tanglefoot®. Vertebrate pests such as squirrels and deer are beigger problems and can be managed as you see fit (netting, angry dog...whatever). Persimmons are related to date plum, (*D. lotus*) and various ebonies, and more distantly to blueberries and other members of the order Ericales.

Stone Fruits (and a Nut)

Stone fruits get their name from having a single central pit, and the term is usually synonymous with the genus *Prunus*. Like apples and citrus, stone fruits have been appreciated for millennia; *Prunus* contains a wide variety of popular foods, including peach, plum, apricot, and cherry. Within the top ten fruits consumed in the United States, the peach/cherry team might get clobbered by the pome (apple/pear) and citrus (orange/grapefruit) teams, but *Prunus* has something the others can't claim: a tree nut. And not just any nut, but the number-one tree nut by consumption: the almond. Pretty good for a single genus.

Unlike apples and citrus, some important *Prunus* species are indigenous to North America, which is a plus for fans of native plants. The big downside of stone fruits is that they may fruit even less reliably than apples and citrus and produce fruit that does not keep well. This may not be a problem if you can deliver to market some heirloom peaches that might otherwise be unavailable; the trick is in getting those peaches in the first place. One of the challenges of *Prunus* spp. is that many bloom early—so early that late frost can wipe out the entire season's potential crop. You need to have enough winter chill to set fruits without encountering blossom-ravaging spring frosts. As a result, the commercial industry tends to be in the southern half of each *Prunus* species's range. Peaches are generally hardy in zones 5–9, for example, but commercial production is concentrated in the warmer half of zone 7 (7b) through zone 9. That's not to say you can't grow a peach in zones 5 and 6—after all, it wouldn't have evolved that cold tolerance if it couldn't set fruit—but harvests would not be as reliable in the colder half.

One trick for urban farmers in the colder half of each range is to trellis the tree against a wall (creating an espalier), which can provide some protection against frosts. Just make sure that it's not a south-facing wall. That may seem counterintuitive, since the sun makes the south wall the warmest. The proscription stems from the fact that the extra heat would encourage the tree to bloom even earlier, offsetting the benefit of growing it near a wall. You want to have extra warmth from a wall while still having the tree bloom at the normal time.

Stone fruits share an affinity for moist, slightly acidic (6.2 to neutral), well-drained soils. Apricots, peaches, nectarines, and cherries favor relatively sandy soil high in organic matter while most plums (especially European plums, but not beach plums) tend to tolerate heavier clay soils. Pretty much any commercially important stone fruit will grow in zones 6–7. Many are hardy down to as cold as zone 4 and as warm as zone 9 (see specific fruits following).

As with citrus, the two options for containers are genetically dwarf varieties or standard-size trees grafted to a dwarfing rootstock. Inherently small species include beach plum (*P. maritima*, about 6 feet [2 m], native), Western sand cherry (*P. besseyi*, about 6 feet [2 m], native), and Nanking cherry (*P. tomentosa*, about 10 feet [3 m] or less, nonnative). In addition, there are genetically dwarf-sized cultivars of usually large species, including 'Compact Stella' and 'Starkcrimson' sweet cherry, 'Northstar' sour cherry, 'Green Leaf' and 'Bonanza' peach, 'Garden Delight' and 'Necta Zee' nectarine, 'Garden Annie' and 'Stark Golden Glo' apricot, and 'Garden Prince' almond. All are fruitful and likely to grow no taller than you are, except for the cherries and almonds, which grow to about 10 to 12 feet (3 to a little more than 3.5 m).

As *Prunus* spp. are closely related, there are only a few dwarfing rootstocks needed to capture most of the commercially significant species. For cherries, the 'Gisela 6' and 'Colt' rootstocks provide semi-dwarfing, while 'Gisela 5' provides full dwarfing (you can keep it under 10 [3 m]). For plums, peaches, nectarines, and apricots, 'St. Julien A' provides semidwarfing, and 'Pixy' full dwarfing. Apricots, plums, and (many) almonds can be semidwarfed with 'Marianna 26–24,' which also tolerates wet soil better than some other rootstocks. There are other rootstocks chosen to produce different-sized trees, to resist particular pests, or to adapt to particular soils. The aforementioned rootstocks are all chosen for one main reason: the presumed need for dwarfing in city living. Depending upon where you live, it might be worth sacrificing a little dwarfiness for some other benefit. Local nurseries or a cooperative extension should be able to help you. Note that even if you have a genetic dwarf stone-fruit tree, you will likely need to prune the roots periodically because they are vigorous plants.

Plums

Plums are among the easiest of stone fruits to grow, at least according to some. On the other hand, plums take up two spots (American and European) out of the five highest maintenance fruits and nuts listed by Purdue University Cooperative Extension Service for edible landscaping. Irrespective of difficulty, you have three main choices. Asian plums (*P. salicina*) bloom early; have large, juicy fruit; and are clingstone, meaning that the pit is attached to the flesh. They are the kind most typically found in supermarkets. European plums (*P.* × *domestica*) bloom late; have medium, drier fruit; and are usually freestone, meaning that the pit can be pulled easily from the flesh. They are the kind of plums dried into prunes and most often turned into preserves. American plums (various spp.) are a third option, though not one that is widely practiced. They are seldom sold on their own, and there's a chance that the only American plum you ever ate was just some genetic contribution to an Asian/American hybrid devised to be extra hardy.

Most plums need to be cross-pollinated; the difference in bloom times (and possible incompatibility) means that Asian and European plums will not pollinate each other. It's good, therefore, to have more than one of each kind. 'Methley' is a (rare) self-fertile Asian plum that is also a good pollenizer (providing pollen to fertilize other cultivars), with 'Stanley' filling a similar role among European plums. Native American plums and Asian/American hybrids can also cross-pollinate Asian varieties. Asian plums are hardy in zones 6–9 and grow better in the South. European plums are hardy in zones 4–7 and are better in the North or anywhere prone to late-spring frosts. Asian/American hybrids are similar to Asian plums but extend the hardiness down to zones 4 or even 3; 'Toka' is a popular, self-fruitful variety hardy in zones 3–8. You can also find trees with two or more Asian or European cultivars grafted onto a single rootstock to ensure pollination. There are several native plums (in addition to beach plums) worth trying in harder-to-satisfy climates, and there are "wilder" European varieties, such as damsons and greengages, worth exploring. European plums are more forgiving of moister soil and partial shade than Asian varieties. They are also left on the tree to ripen, while Asian varieties are picked slightly before full ripening.

Cherries

Cherries generally hue to the plum's cultural conditions, although cherry trees are larger. We're used to sweet cherries (as in "life's not a bowl of," *P. avium*), although there are also sour (or tart) cherries used mainly for baked goods. Sweet cherries are generally hardy from zones 5–9 (though best in 5–7) and require cross-pollination by another cultivar; two self-fertile exceptions are 'Stella' and 'Starkcrimson' (which also happens to be a genetic dwarf). Sour cherries are hardy from about zones 4–9 and self-fertile; a popular variety is 'Montmorency." (Remember, the most popular fruits are seldom the easiest, so 'Starkcrimson' and 'Northstar' may be worth trying first). Sour cherries are also, by some accounts, the least troubled stone fruit in terms of pests and diseases. Other relatively low-maintenance options are Nanking cherry (*P. tomentosa*) or Western sand cherry (*P. besseyi*). Neither is very susceptible to pests, and Nanking cherry also tolerates some shade. Named cultivars of both exist—and would probably be your best bets—though they are usually sold as unnamed species.

Peaches and Nectarines

Peaches generally are hardy in zones 4 or 5–9, and nectarines—which are essentially weak, bald, but delicious peaches—are hardy in zones 5–8. Most peaches and nectarines are self-fertile, needing no pollinator, with some notable exceptions among peaches being 'J.H. Hale,' 'Hal-Berta,' 'Candoka,' and 'Mikado.' Among the normal self-fertile kinds, 'Reliance' is a common peach recommendation, and 'Red Gold' is common for nectarines. 'Reliance' has somewhat smaller fruits for a peach but is among the cold hardiest and is reliable throughout its range (about zones 4–8). 'Red Gold' is the standard nectarine but is somewhat susceptible to disease.

Apricots

Apricots are treated much like peaches and nectarines, and there are varieties adapted to zones 4–9. Most are self-fruitful, with 'Rival,' 'Perfection,' and 'Riland' being notable exceptions. 'Moorpark' is a classic cultivar recommendation, although 'Harcot' may better suit those in the northern half of the apricot's cultural range. Most of the apricots grown in the United States are European cultivars (*P. armeniaca*), although the fruit is originally from Asia and has many other varieties. Manchurian apricot varieties (*P. armeniaca* var. *mandshurica*) are among the hardiest; 'Mandan' is a variety developed by the USDA in the Great Plains that is particularly hardy. Japanese apricot (*P. mume*) is somewhat less hardy (down to about zone 6) and bears fruit that—even if little appreciated here—are widely popular pickled in Japan (umeboshi).

Almonds

Almonds also require similar cultural conditions to peaches but have a narrower range—about zones 6–9, although 7–9 would be a safer bet. Not only do they share a preference for mild winters but they also like long, dry summers. Commercial production, not surprisingly, is very concentrated in just a few locations. The two best varieties for urban farmers are likely 'All-in-One' (semidwarf) and 'Garden Prince' (genetic dwarf), both self-fertile cultivars developed by Zaiger Genetics (through traditional breeding not genetic engineering). 'All-in-One' is very cold tolerant (likes 500 or more winter chill hours), making it good for the northern part of the range, but it requires a hot summer. 'Garden Prince' has a low chill requirement (about 250 hours) and is better suited to the more southerly of the almond range. **Note:** Do not plant almonds near other stone fruit because of the risk of getting bad-tasting fruits if they cross-pollinate.

Companions, Relatives, and Pests

Get yourself a drink and sit down. It's bad. If we were to focus just on big diseases common to all or most of the *Prunus* spp., you'd have bacterial canker, bacterial spot, black knot, brown rot, peach leaf curl, and powdery mildew. Invertebrate pests would include borers of various stripe, scale insects, and (in the East) the dreaded plum curculio. The best general prophylactic, of course, is to choose resistant varieties, plant in the right conditions, and maintain the plants' health vigilantly. Avoid shallow soils, which increase susceptibility to bacterial canker.

Bacterial canker appears mainly on woody parts of the tree as lesions. Cut them out with a few inches of undiseased tree to spare, and cauterize the wound with a blowtorch. Avoid pruning trees until late winter. The University of California Statewide Integrated Pest Management (IPM) Program also recommends fall or spring foliar feedings of complete micronutrients. Somewhat hardcore options include fungicides, such as Nordox 75 WG, or products with competing microorganisms, such as *Bacillus subtilis* QST 713.

Bacterial leaf spot appears as dark freckles on leaves and fruit. It occurs in humid conditions, especially on susceptible cultivars. Small outbreaks don't warrant treatment, and good cultural practices such as removing dead and infected foliage and watering early in the day (to allow for drying) and at ground level (instead of on leaves) help prevent spread of the condition. Conventional practice is an early application of copper-based fungicide before symptoms appear, followed by weekly applications of antibiotic. But that's a little drastic for small-scale urban farmers, who might opt for Nordox 75 WG or *B. subtilis* QST 713 if they really feel the need to go the chemical route.

Reddish, puckered new growth and contorted shoots suggest peach leaf curl, a fungal disease. If it affects your peaches or nectarines this season, the only effective treatment is a preventive one to help next season's crop: apply a fungicide such as a Bordeaux mixture after the leaves fall in autumn.

Powdery mildew, which appears like a thin film of white or grey velvet, is another fungal condition that develops under similar moist, warm conditions. Good cultural practices limit the risk, and, unlike with peach leaf curl, it can be controlled after an outbreak appears. In addition to heavy-duty synthetic fungicides, there are OMRI-listed products with less toxic ingredients such as jojoba oil, rosemary oil, kaolin clay, copper, and various competing organisms. If powdery mildew poses a big problem one season, preventive fungicides may be a good idea for the next. You can find various recipes for baking-soda-based sprays online.

Brown rot is the bane of many *Prunus* spp., appearing first as soft brown spots on fruit. The fruit can then develop velvety patches and progress into a shriveled shell called a mummy. Getting rid of the mummies and all infected plant parts and following good watering practices may well control brown rot. If it has posed a serious problem in the past, preventive sprays—many of the same ones applicable for powdery mildew—may be a good idea early in the season.

Another fungal disease, black knot, infects mainly plums and cherries. It appears as black, tumorlike growths on the woody parts of the plant. As with bacterial canker, these knots should be cut out with several inches to spare. Preventive fungicides may help minimize spread, but look around your farmstead: if there are nearby infected wild or ornamental trees, it's probably best to avoid plums and cherries altogether. 'Stanley' is especially vulnerable, while 'Methley' has moderate resistance.

There is a wide variety of borers with a wide variety of appetites, but most can be controlled with just a few cultural practices. Whitewashing trunks with a 50/50 mixture of non-enamel white interior latex paint and water reduces sunburn damage, which would invite borers. Sprays containing garlic extract may repel many borers, while those containing spinosad kill many (if not all) of them. You can find OMRI-listed products containing both substances.

Plum curculios, which look like tiny anteaters, can be controlled in the same way as you would for apples, and scale insects in the same way as for citrus plants.

Like apples and pears, *Prunus* spp. are members of the rose family. Cherry laurels (*P. laurocerasus*), popular landscape plants, are in the same genus. It's a good idea to keep *Prunus* fruit trees away from roses if the roses frequently host powdery mildew or other fungal diseases.

CHAPTER 8

Plant Management

Once you have an idea of what kinds of plants you want, the next question is: how do you take care of them? As with dogs or cats, you can't just pick out the plants you want and expect them to take care of themselves. They'd probably die. And even if (unlike with dogs or cats) you didn't go to jail for it, you wouldn't be a very good urban farmer. If you know the basics, however, you can decide in advance just how much of a commitment you want to make to your future meal and plan accordingly.

The most natural way to grow most plants is to start with seeds. Buying "seed starts" or young plants is easier, but it limits your selection and is more expensive. You should try at least some seeds, at least once.

Starting Seeds

Almost invariably, gardening texts refer to growing a plant from seed as "rewarding." And it is rewarding, and amazing, and potentially economical. On the other hand, few things are as frustrating as having the seedling you've tended for weeks succumb to damping off or shrivel into a tiny heap when you forget to water it for one day. As with your cultivation of plants, generally fortune favors the prepared. And the first choices for the prepared are the container and medium.

Containers

Most kinds of containers will do, especially plastic ones such as nursery trays or yogurt cups. The main requirements of seed-starting containers are that they're about 3 inches (7.5 cm) tall and sterile. If you're using impermeable ones, such as plastic, disinfect them using a 10 percent bleach solution, rinse, and dry before adding the soil. You can also try expandable peat or coir pots, perforated rock-wool cubes the size of dice, fillable peat pots, or little fillable pots made out of newspaper with a tool just for that purpose. The great virtue of these options—all permeable and (except for rock-wood) biodegradeable—is that when it comes time to transplant, you can do so without disturbing the seedling's roots. Make sure that the containers drain well (either through holes or natural capacity, like peat pots), and place them on a shallow-lipped tray (such as an upside-down plastic coffee-can lid) to catch excess water. Hollow containers should be about three-quarters full of the seed-starting medium.

Oh, and one thing I should have mentioned before: label the containers. Seedlings tend to look amazingly alike.

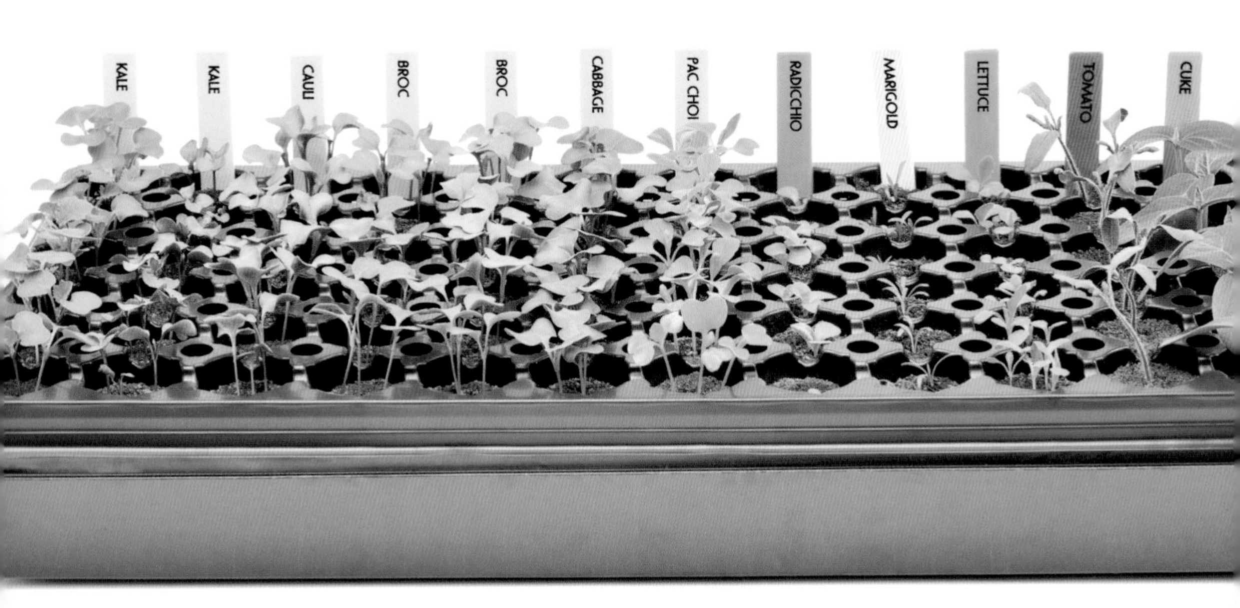

Starting seeds near a window is not ideal, but it works!

Growing Medium

Seed-starting is one area in which even many die-hard organic types insist on sterile soil. Your seedlings may not thrive the way they could with beneficial microorganisms about, but they won't be as prone to the ravages of bad microorganisms, which seem to manifest an almost sentient thrill in killing seedlings. You can buy sterile seed-starting mix, or you can create your own (see chapter 5). These mixes are typically finer than other media so that seeds don't settle down too deep.

Generally speaking, seeds are planted at a depth of two to three times their diameter (so a 1-millimeter seed would go 2 or 3 millimeters down), prefer slightly warmer than room temperature (60–75 degrees Fahrenheit [15.5 to 24 degrees Celsius] during the day and slightly cooler at night), and need to be kept moist (but not waterlogged) at all times. However, whatever your seed packets advise trumps these rules of thumb. Seed containers are usually covered with some kind of plastic hood (even if just tented plastic wrap or a plastic bag) to maintain humidity. Here's where the sterility pays off, since these warm, wet conditions are also perfect for

Building a Community around Food

Rick Bayless is not so much a celebrity chef as a force of nature—a good force from the perspective of urban agriculture. Sure, he has celebrated restaurants in Chicago—Frontera Grill, Topolobampo, XOCO, Leña Brava, and Cruz Blanca—as well as Tortas Frontera, Frontera Fresco, and Frontera Cocina restaurants in other locations. Yes, his cookbooks are classics, his TV shows are treasured, and he's widely appreciated for his knowledge about traditional Mexican cuisine. Oh, and there's that delicious line of Frontera foods at Whole Foods supermarkets.

Chef Rick Bayless tends to some of the crops grown on top of his restaurant, Frontera Grill, in Chicago.

Despite all this, however, Bayless is rather unassuming—a "reluctant rock star" according to the *Houston Press*. And he is committed to sustainability. In summertime, EarthBoxes full of heirloom peppers and tomatoes grow on the roof of Frontera Grill, destined to become salsa. His restaurants also feature the tomatoes, chiles, beans, and herbs grown in the 1,000-square-foot (93-sq-m) rooftop garden over XOCO. The backyard garden at his home in Chicago yields more than 1,000 pounds (454 kg) of vegetables and herbs each year. And in the basement of that home, microgreens reach their peak under plain-old fluorescent bulbs. His restaurants draw many of their other ingredients from small local farmers. He even started the

One view of the garden behind Rick Bayless's home. The roughly 1,000 square feet (93 sq m) devoted to his restaurants grows vast amounts of produce every year—in Chicago!

Frontera Farmer Foundation to support local sustainable farms through small grants.

"When you see what can happen to a community when it celebrates its local agriculture," says Bayless, "it is really amazing, and it is the first step to solving the problems we have with the American diet." He's not kidding about the effect on a community, or about the steps. Bayless's own success comes despite the fact—or more probably *because* of the fact—that he's a real one-day-at-a-time guy. You can even tell by his language. The motto of his restaurants is "to make the right next step," and he refers to sustainability not as a state but as a destination—"the road to sustainability," something measured out in steps.

"Once you start making processed food look tawdry and cheap," he says, "then you can start planting in people's minds the question: 'Really, I'm supposed to eat that stuff?'" That makes fresh, local food look great, and it raises another question: "'Oh, I have to cook it?' Then the next thing you know," says Bayless, "they're not eating so much, because they actually have to invest something in it. And those are direct steps toward solving all of the problems we have now."

The investment yields other rewards, too. "People will spend more time around the table because someone would have made something," he says, "it would change the whole fabric of family life, of community life." Bayless sees these changes already affecting Chicago, with its vibrant community gardens and small farms, culinary emphasis on local foods, and commitment that extends right up the roof of city hall. "That's why I'm so dedicated to developing local agriculture, urban agriculture," he says. "I feel like we can touch more people here in the city with good, fresh local food. And I'm not even talking about how it's grown. I'm just talking about that fact that it's here. That can really change a community."

fungi, such as the damping-off disease that causes tender sprout stems to brown, wither, and flop over. Lighting from a window is OK, but try to protect against large temperature fluctuations near windows in cold regions. Artificial lights are great alone or as supplements to a window (see Lighting in this chapter). In cool spaces such as basements, fluorescent lights may have an edge over other artificial lighting because they provide some heat, taking up the heat a few degrees in daytime (or when turned on) and allowing it to drop back in nighttime (or when turned off).

Once seeds have sprouted, make sure that the growing medium is constantly moist, but also increase the air circulation. Watering with a fine mist is best; if you can't do that, at least protect the tiny seedlings from being knocked down by a stream of water. When seeds have their first true leaves, start fertilizing using a half-strength water-soluble fertilizer and thin them (pinch out seedlings causing crowding) to maximize light and air exposure.

One or two weeks before you intend on transplanting the seedlings outside, you need to harden them off. It's like boot camp for the young plants, which will have hitherto enjoyed a life not unlike a spa—balmy weather, tanning beds, a bottomless supply of drinks. To acclimate them to their new life, you need to introduce them to real sunlight—which is much more intense—and temperatures that are likely much chillier. There's no hard and fast rule for how to do it, but you want to first put the plants out in a shady place at a warm time of day for a few hours, then bring them back in, and then take them out to progressively sunnier situations for a little longer each day. Afternoon is a good time to plant outside when the chosen day arrives.

Lighting

Neutral, white light is actually a combination of different wavelengths, usually thought of as red, green, and blue (like the pixels on your TV or computer screen). Plants, for the most part, only need two of them: red and blue. It seems strange, since plants are green, right? But think about it. Plants appear green to us because when white light hits them, they absorb the red and blue, leaving only green to bounce off and hit our eyes. Weird.

Light for seed-starting sprout production or simple indoor farming can be provided by a south-facing window with good light, artificial light, or some combination of window and artificial light. Incandescent bulbs are inefficient producers of light, as so much energy gets diverted into heat—but that might be desirable if you have a very cool location for growing. Otherwise, the best and cheapest long-term options if you're starting from scratch are fluorescent bulbs and LED (light-emitting diode) lamps.

The standard fluorescent setup involves two 4-foot (1.25-m) tubes in a fixture that holds them a few inches apart and parallel. The conventional recommendation is to use two 40-watt tubes for 14 to 16 hours a day, 6 to 8 inches (about 15 to 20 to cm) above the plants or about half that distance for seed-starting. As the plants grow up, you move the fixture up (or the plant down). This produces about 20 lamp watts per square foot (two 40 watt lamps = 80 watts total; 80 watts divided by 4 feet = 20 watts per foot). Use that as a rule of thumb.

Various trays of crops grow beneath red and blue LED grow lights.

A typical 23-watt screw-in compact fluorescent light—often billed as equivalent to a 100-watt incandescent—could provide equivalent light to cover slightly more than a square foot. But which color should you get—soft white, bright white, cool white, warm white daylight, or full spectrum? Full spectrum might be best overall, but if you're using more than one bulb, it may be as effective (and certainly cheaper) to mix bulbs from opposite ends of the spectrum: half leaning toward blue (daylight) and half

toward red (soft white or warm white). Those are the colors that plants absorb most for photosynthesis. Cool white and bright white tend to be in between—greens and yellows—which plants don't really use. If you're supplementing light from a window (which leans blue), choose soft or warm white.

However, the clear winners are LED lights, which are roughly as superior to fluorescent lights in terms of lifespan and energy efficiency as fluorescent lights are to incandescent bulbs. The price of LED bulbs have dropped dramatically, and although they are still more expensive at the outset, they consume a fraction of the energy of other types of lights and last for tens of thousands of hours, so you'll recoup the extra cost before you know it. Since they channel almost all energy into light (rather than heat), another advantage is that they are very cool. A plant (or person) can touch them without getting burned. Finally, LEDs can be calibrated to very specific wavelengths that can be beneficial for plants. In fact, it is easy to find LED arrays (bulbs, panels, or "UFO"-style round units) fine-tuned to plants' needs, such as vegetative growth (more bluish) or flowering and fruiting (more reddish).

Note that while you can buy LED panels that include only red and blue—channeling all of their energy into the wavelengths that plants care about—think twice before using them in your living space. Even bright red/blue grow lights appear dim because our eyes are only half as sensitive to those colors as they are to greens, and the purplish glow they emit is kind of freaky.

While it's best to start with LED lights if you can, you can always swap them in gradually. For example, you can now buy LED replacements for fluorescent tubes.

On Sprouts and Microgreens

Although starting seeds may be only the first step—if not one skipped entirely—by most urban farmers, the endeavor can be in itself very nutritious, profitable, and forgiving of less-than-ideal conditions. If you have a large expanse of flat areas (such as shelves or tables) and good artificial or natural light, one option is to start far more seeds than you would plant yourself and sell the excess to other farmers unwilling or unable to start seeds themselves. Another option is to grow tiny vegetables—the realm of sprouts, (micro)shoots, and microgreens.

Sprouts are the fastest of the three and are basically seeds that have sprouted, begun rooting, and put out their cotyledons, the so-called "seed leaves" that emerge before regular leaves, which look different. Bean sprouts and alfalfa sprouts fall into the sprout category, although many other seeds, such as broccoli, radish, sesame, and onion, can be sprouted. The whole shebang is eaten. For the most part, they don't even require light. Now that's low maintenance.

Unlike sprouts, which are basically young roots, shoots are primarily the young above- ground portion of a given plant—the tender stem, leaves, and other appendages. Shoots are germinated like sprouts and then put in a medium such as soil, vermiculite, or a textile to root. Popular shoots include corn, wheatgrass, and peas (pea shoots are sometimes called *tendrils*). When harvested, they are cut just above the medium, so you get lots of little leaves but no roots.

Microgreens are essentially just baby versions of other plants we eat as greens, such as arugula, lettuces, mustard, and sometimes herbs. Microgreens are grown and harvested in the same way as shoots and take a month or less to mature, depending upon the particular plant. Unlike sprouts, both microshoots and microgreens need light.

Under the right conditions—which are not demanding—you can harvest sprouts every few days, microshoots every two weeks or so, and microgreens every two to four weeks. These qualities make them very appealing to urban farmers. For example, Wally Satzewich of Wally's Urban Market Garden in Saskatoon is a big fan. Chef Rick Bayless grows microgreens for his restaurants in the basement of his Chicago home, favoring spicy brassicas, colorful plants such as 'Bull's Blood' beet, and chrysanthemum greens (*shungiku*). He's even found that many of the microgreens prefer the artificial lights of the basement to a greenhouse.

If you have the basic infrastructure, sprouts, microshoots, and microgreens are also tremendously inexpensive: you can buy a pound of organic seeds for just a few dollars. A tablespoon of seeds will easily produce a pint of sprouts. It's important to buy seeds specifically for sprouting, however, to avoid any soil-borne contaminants.

Creating a Garden Calendar

For everyone but the most casual of recreational farmers, a garden calendar is a must. Keeping track of when to plant and harvest, it ensures that you avoid rookie mistakes (planting lettuce in the height of summer), maximize yield (by planting hot-weather crops in the spaces just vacated by your cool-weather harvest), and stagger your harvest over a long period (instead of a one-week super-harvest amid twelve weeks of nothing).

You'd think that, for something so useful, there would be plenty of templates and other software resources available, but if there are, I haven't found them. Nevertheless, it's pretty easy to do it yourself with graph paper, an actual calendar, or a spreadsheet. I like the spreadsheet approach because you can easily copy the previous year's version, correcting for whatever you learned. Though a spreadsheet

Crop	JAN		FEB		MAR		APR		MAY	
	early	late	early	late	early	late	early	late	early	late
Broccoli 'De Cicco'						start seeds indoors	→	transplant	→	→
Lettuce 'Cosmo'					start seeds indoors	→	transplant	thin	harvest	
						start seeds indoors	→	transplant	thin	harvest
							direct sow seeds	→	thin	→
								direct sow seeds	→	thin
									direct sow seeds	→
Tomato 'Mountain Princess' (Determinate)					start seeds indoors	→	→	transplant	→	→
						start seeds indoors	→	→	transplant	→
Tomato 'Black Brandywine' (Indeterminate)					start seeds indoors	→	→	transplant	→	→

is what's used in the accompanying illustration, you can easily do the same with graph paper or with a paper calendar. (Of course, there's a very good chance that your local cooperative extension already has a planting calendar tailored to your frost dates and filled with locally adapted cultivars. Check it out.)

Once you have your calendar ready, the first thing to do is to mark down the average dates of the last spring frost and first autumn frost in your area. For me, that's around April 24 and October 17. As you can see in the illustration, I've divided each month into "early" and "late," each comprising about two weeks. I've highlighted the last spring frost in green in late April and the first autumn frost in blue in late October. I've marked down seeding, transplanting, and harvest dates, all based on actual cultivars sold by the Southern Exposure Seed Exchange.

JUN		JUL		AUG		SEP		OCT		NOV		DEC	
early	late	early	late	early	late	early	late	early	late	early	late	early	late
harvest			direct sow seeds	→	→	→	→	harvest					
					direct sow seeds	→	thin	→	harvest				
						direct sow seeds	→	thin	protect	harvest			
harvest													
→	harvest												
thin	→	harvest											
→	harvest												
→	→	harvest											
→	→	→	harvest	harvest	harvest	harvest	harvest	harvest	harvest				

First is broccoli 'De Cicco.' Since broccoli is primarily a cool-weather crop, I aim for spring and fall plantings. The description suggests that it takes about twenty-five days to get to transplant size. That's about four weeks, so I aim to start seeds indoors four weeks before late April. That should produce good-size transplants by the approximate last frost date when I intend to plant. 'De Cicco' takes about forty-nine days (seven weeks) to grow to harvest size, so I mark down the expected date in early June (two weeks in June plus four weeks in May plus one week in late April, for a total of seven weeks). Southern Exposure says to plant fall crops ten to twelve weeks before the first hard-freeze date, so I assume a harvest just before the first autumn freeze and count back twelve weeks (six squares) to a date to directly sow in the ground.

Lettuce is another cool-weather crop, so I likewise aim for spring and fall plantings. Since I want a nearly continuous supply of lettuce greens, I stagger five plantings for spring harvest and two for fall harvest. I know from the catalog that the 'Cosmo' lettuce cultivar takes about fifty-five days (roughly eight weeks) to reach full size from direct sowing and about thirty-five to forty days from transplants (which tells me that it takes about fifteen to twenty days—two to three weeks—to produce a transplant). I then strategize direct sowing and transplanting based on these numbers, avoiding the hottest part of the summer.

I follow the same timing strategies for tomatoes, planting two cultivars. 'Mountain Princess' is an early-ish tomato—fruiting about sixty-eight days from transplant—and is determinate, meaning that the tomatoes will all ripen at about the same time. I've staggered two plantings of 'Mountain Princess' so that I can start enjoying tomatoes as early as late June. 'Black Brandywine' needs considerably more time—eighty-five days from transplant—and is indeterminate, meaning that it bears fruit continuously. Combining the cultivars in this way offers the possibility—if not the promise—of a long yield of tomatoes.

This combination of vegetables should, ideally, produce a near continuous harvest of at least *something* for almost half a year. That may not happen, of course—there are all kinds of animal, mineral, vegetable, and meteorological bugbears waiting to stomp on your hard work—but success is far more likely to happen if you're prepared in this way. You can bring even more precision into the mix by dividing each month into four weeks or even individual days, but then you'll have a very large spreadsheet. It also makes sense to stagger "same time" plantings just a bit. With broccoli 'De Cicco,' for example, the calendar makes it seem as if all of the broccoli seeds are sown at the same time and all of the plants transplanted at the same time. In fact, it would make more sense to make two or three sowings just a few days apart in late March and to transplant them out just a few days apart in late April. This will help extend the harvestable period to the planned two-week window as well as save some of your crop in the event of an unexpected cold snap.

Another nice use of the calendar is to plan other farming activities, particularly during off periods, such as putting down organic phosphorous/potassium fertilizers in early November (and organic nitrogen ones in early March) or disinfecting tools and testing germination of last year's seeds in January.

Maximizing Production

One of the major challenges of urban agriculture is space. You don't have a lot of it, so you want to make the most of what you have. But it may be potentially more abundant than you think. SPIN-Farming aims for urban growers to gross $50,000 per year with just a half acre, while the GROW BIOINTENSIVE sustainable farming methods offer the potential to grow enough food to feed a person for one year on just 4,000 square feet (372 sq m)—less than a tenth of an acre. You might not have the space (or commitment) to take on one of these programs, but there are still ways you can maximize production.

Intensive Planting

A good way to grow more plants is to simply grow more plants—intelligently. A common way to do this recommended by many intensification-focused farmers is to plant more closely together than seed packets or traditional gardening books would advise. There are limits, of course, but if you space the plants well, you will lose very little productivity from any given crop and gain weed suppression and water conservation through shading. (But remember: more plants mean more nutrients and water needed.)

Raised Beds

Raised beds make intensive planting easier since you can ensure several inches of quality soil, use grids or twine to measure out distances precisely, and take advantage of the season-lengthening effects of raised beds—even more so if you build simple hoop houses on top of them.

Mulching

Mulching also aids in weed suppression and water conservation while adding in a helpful amount of organic matter. It also helps to prevent dust from getting kicked up, which is helpful if you're growing in an area with known contamination.

Going Vertical

Intensive planting often makes efficient use of vertical spaces. Vertical planting can be done using stakes in raised beds or containers, growth along fences, fruit trees trellised along walls, hanging baskets...you name it. There are now so many resources available to stack or hang planters, or grow plants up a wall or fence that some version of vertical production is definitely worth a try.

Interplanting

Interplanting, which involves growing plants in close proximity, takes intensive planting one step further. (It is sometimes called intercropping or multicropping, though not uniformly.) Interplanting lettuce with radishes is a perfect example. Radishes grow quickly, partially screening lettuce, which needs some shade. In addition, radishes have relatively deep roots and lettuce has shallow ones, so they're not in close competition for water or nutrients. You also harvest them at different times, and can even interplant another longer-season plant such as pepper or broccoli with them.

Succession and Relay Planting

Succession planting and relay planting are employed to yield two or more crops in the same space over the course of a season. As with interplanting, there is some inconsistency in the use of the terms. Some use *succession planting* as a term for successive sowings of the same plant and relay planting to refer to successive sowings of different plants, while others treat them vice versa. Either way, the concept is the same. Your garden real estate is valuable, so why give six months' worth of a given space to a crop that will only live there for one or two?

A typical succession planting would be a cool-season, early-yielding crop such as peas or lettuce followed by a longer, warm-season crop such as beans or tomatoes. Often another cool-season, short-term vegetable is possible as a third crop. Note that there can be a significant time difference in the days to harvest between cultivars of a given vegetable. The Southern Exposure Seed Exchange, for example, offers a "Sophie's Choice" tomato yielding in fifty-five days and a "Martian Giant Slicer" yielding in ninety-five. That's half a summer between them. If you plan carefully, live in a milder climate, focus on short-season or "baby" crops—which are harvested while they are still tender and young—or simply have an insatiable appetite for radishes, it is possible to have even more harvests.

Cool-weather crops typically include lettuce, spinach, radishes, peas, broccoli, mustards, and onions. Warm-season crops typically include beans, turnips, corn, tomatoes, and peppers. As with so much in agriculture, there are no hard-and- fast rules. You may be able to harvest cool-weather crops in the hottest part of your summer, and warm-weather ones up to the first freeze in autumn, and you may find cultivars against type. The key is to do your research before buying.

Polyculture

If the primary aim of intensive planting is to increase the quantity of a yield through a kind of spatiotemporal harmony, then polyculture might be seen as

Two woven wooden plant protectors cover organic cabbages. Orange marigolds act as companion plants to deter pests.

aiming to increase the quantity or quality of that yield through smart mixed plantings of particular plants. The two are not mutually exclusive—you could intensively plant a polyculture, for example—but polyculture generally refers to something more nuanced.

Oats and peas constitute a basic polyculture commonly employed as a cover crop. The oats serve as trellises for the peas, and the peas increase the yield of oats because of their nitrogen-fixing properties. The peas also kick up the protein content of the harvest, which helps improve the blend for its usual purpose: livestock fodder. People have been polyculturing crops for perhaps as long as agriculture has existed. Not only is it more interesting but it also can be more sustainable for the soil, enable a higher density of production, and offer a more complete nutritional profile for the ultimate consumer.

The simplest form of polyculture is companion planting, which could involve something obvious, such as peas and oats; something less obvious, such as planting marigolds to repel harmful nematodes; or something believed effective for unknown reasons, such as planting tomatoes and basil together. There are many reasons why a particular companion planting may or may not work. One might attract bugs beneficial to the other—such as ladybugs, bees, and predatory/parasitic insects—or repel harmful ones. One might even fall on its own sword and be a trap crop that attracts all the bad bugs. A deep-rooted species might bring up water or nutrients that its companion's roots can't access or fix atmospheric nitrogen into the soil. Other proposed mechanisms include the biochemical, the astronomical, or even—as some would say of biodynamics—the supernatural.

Companion planting is not infallible, however. Science has confirmed some useful relationships, debunked others, and reserved judgment on many. Perhaps the most popular example of traditional companions supported by science is the mound planting of beans, corn, and squash—the "three sisters"—common to many Native American peoples. For such a simple arrangement, this combination provides complex and powerful interactions. Broad, scrambling squash leaves help suppress weeds and conserve moisture for all three sisters. Its stems also sport spiny stems that may deter some pests. The cornstalk serves as a scaffold for the beans, its root exudates nourish the beans' rhyzobia bacteria, and its leaves cool the beans through transpiration. Beans fix nitrogen in the soil, diminishing its competition with the other two and boosting next year's fertility—a particular boon for corn, a heavy nitrogen feeder. They also help anchor cornstalks against the effects of wind.

A disease or insect pest that would devastate one sister would likely spare the others since they are not closely related. Forget being of different species and genera—they are from different orders, which is roughly the same level of difference as between humans and hamsters. They are even meteorologically complementary, as homesteader John Vivian has observed. Squash and beans will tolerate partial shade that hobbles corn production, while established corn can better withstand the extremes of drought and flooding. Beans can overcome the twin scourges of high winds and low fertility. Corn can even survive the complete absence of bees, since it is pollinated by wind. The combination of plants is also a healthy one, to boot, for humans and hamsters alike. Neither the beans nor the corn possesses a full complement of amino acids, but together they provide a complete protein. Squash throws in a heap of beta carotene, potassium, and other nutrients.

A "three sisters" planting at the USDA People's Garden early in the growing season.

The three sisters illustrate companion planting par excellence as well as a transition from simple companion planting to what permaculturists would call "polyculture." This is particularly true when considering that traditional regional variations incorporate a fourth, fifth, or even sixth sister. Many tribes in the Northeast, High Plains, and Southwest planted sunflowers along the borders of three-sisters plots (always along the northern borders to prevent shadowing the others). Spiderflower (*Cleome serrulata*) was apparently included in some southwestern plantings, and some tribes sometimes included quinoa or amaranth—two related, strikingly nutritious annuals. Long-blooming sunflowers and spiderflowers send bees into a tizzy—hence many of *C. serrulata*'s other common names, including Rocky Mountain bee plant, bee spiderplant, and bee weed. Both attract pollinators for beans and especially squash, whose blossoms bloom just briefly. Sunflowers, quinoa, and amaranth abound with nutritious seeds that attract birds. Birds, in turn, can eat insect pests, pollinate some flowers, and leave what permaculturist Toby Hemenway calls "small gifts of rich manure."

Once we get to a fourth sister—maybe even the first three are enough—we move to what permaculturists would call a "guild," defined by Hemenway as "a group of plants and animals harmoniously interwoven into a pattern of mutual support, often centered around one major species, that benefits humans while creating habitat." The three sisters engage plants and bacteria into mutually beneficial relationships, while a fourth (or fifth or sixth) brings in animals. It is an arrangement that emulates the messy polyculture of natural ecosystems rather than trying to oppose it at every turn in shape (neat, rectangular fields), composition (vast monocultural plantings), and self-sufficiency (drenching of herbicide, pesticide, and synthetic fertilizers). The essence of the guild, according to Hemenway, is not so much the number of members, but the number of "functional connections" between them.

Planning a guild might be too ambitious for an urban farmer, but the ecological thinking it involves could inform your choice of plants and how they could work together. If you have a small backyard, for example, you might plant an Actinidia vine along the east or west wall to insulate the building, provide fruit for you and birds, and reduce rodent pressure by attracting cats. You could pot culinary herbs and flowers known to bloom at different times to serve as insectaries for beneficial bugs. Sunflowers could help remedy a compacted, weedy corner by breaking up the soil with strong roots and suppressing weeds with shade and the allelopathic chemicals they produce, which are toxic to other plants. They also attract pollinators, beneficial predator insects, and birds. When you begin to think not just of particular plants but of functions they serve—and the proximity of different crops both in space and time—you will begin to weave together the benefits of intensive planting and polyculture.

Fertilizing

All plants need nutrition, and that means fertilizer of some sort. As mentioned, packaged fertilizers include a three-number sequence detailing the amount of the macronutrients that plants need to survive: nitrogen (N), phosphorous (P), and potassium (K). Alfalfa meal, for example, might be labeled 5-1-2, with each number indicating the percentage of available nitrogen (5 percent), phosphorous (1 percent), and potassium (2 percent). What's the other 92 percent? It depends on the fertilizer. In alfalfa meal, most of it is carbon. Bonemeal is nearly one-quarter calcium, for example, and granite dust is about two-thirds silica. The non-NPK part of fertilizers is known as the ballast, and a given ballast can be good, bad, or neutral for plants. Many fertilizers also include trace elements or micronutrients in their ballasts, or even as the headliners. Kelp meal is virtually 0-0-0, for example, but contains micronutrients hard to find elsewhere.

Some high-nitrogen, highly soluble fertilizers release nitrogen quickly and can "burn" the plant if overapplied, inhibiting the plant's ability to take up enough water and leaving scorch marks on the leaves. Heavy watering can sometimes remedy this. Although such "hot" fertilizers are often synthetic, uncomposted animal manure can have the same effect, with dry rabbit manure sometimes considered an exception. Composted animal manures, other organic amendments, and specially formulated synthetics release their nutrients slowly.

Now the tricky math part. Some gardening guides will suggest application of a particular macronutrient (often nitrogen) at a particular rate (such as 1 pound [0.45 km] per square yard [just shy of 1 sq m]). To accomplish this with alfalfa meal, you'd need to apply 20 pounds (9 kg) per square yard/meter because only 5 percent of that (or 1 pound [0.45 km]) is available nitrogen. You'd only need 10 pounds (4.5 kg) of blood meal, however, since that is 10-0-0, or 10 percent available nitrogen.

What to Use

The choice of how to meet a requirement entails a lot of considerations, including cost, the availability of options, and whether or not other macronutrients are required. If you're going the chemical route, the math is fairly straightforward. Organic/slow-acting fertilizers are a little more difficult to calculate if you're making them yourself, but luckily you can find an application from the Oregon State University Extension Service to calculate quantities for a wide range of ingredients (see Notes & Resources). Either way, aim for restraint. Slight underfertilization might reduce your crop yields, but overfertilization can kill your plants. Overfertilization can also cause nitrogen pollution of waterways, a major environmental problem.

In the broadest terms, you can think of the macronutrients as reflective of the plants themselves. Nitrogen fosters green, leafy growth; that's why, for example, new-lawn mixes are disproportionately high in nitrogen, and homemade tonics for greening up plants (such as those involving beer, ammonia, or urine) are naturally high in nitrogen. That's also why you stop or reduce nitrogen supplementation when an edible plant is fruiting—we want the energy to go into the future meal, not into leafy growth.

Phosphorous is root food. That's why gardeners apply high-phosphorous fertilizers at planting and just after the blooming of lilies, gladioluses, and other flowers that

Phosphorous-based fertilizers vary widely in their available phosphate, so be aware of which kind you're buying.

grow from bulbs—to coddle the all-important underground storage structures, be they bulbs, corms, rhizomes, or tubers. Phosphorous is also vital for flowering, fruiting, and setting seed, making it popular in fertilizers for vegetables and fruiting trees.

It would be wonderfully parallel to say that potassium makes stems grow, but that's not the case. In fact, potassium does not have a claim on any particular part of plants but plays a supporting role in many processes, including photosynthesis and water management.

If you're planting in the ground, ideally you will get the soil analyzed before engaging in any significant fertilization. If you're not that much of a forward thinker and have already installed some plants in the ground, you can look out for symptoms indicating deficiencies in particular nutrients. It's imprecise but better than nothing. Some symptoms of macronutrient problems include:

This chart is just an example of problem indicators that could be related to nitrogen, phosphorous, and potassium; many other problems may produce the same specific symptoms. Fusarium wilt might cause yellowing and lack of vigor like a

	TOO LITTLE (DEFICIENCY)	TOO MUCH (TOXICITY)
NITROGEN (N)	Spindly growth; yellowing of leaves, with older leaves paler; high sensitivity to water stress. Occasionally greener leaves with yellowish or reddish veins.	Rampant green, leafy growth at the expense of flowers and fruits. Soft, wilty stems.
PHOSPHOROUS (P)	Slow growth on otherwise dark-green, healthy-looking plants. Poor flowers and fruit. Occasional purplish/reddish tinge to stems, leafstalks, and leaf undersides, and/or bluish cast to leaves.	Phosphorus toxicity is extremely unlikely in soil-grown plants. No telltale signs indicate its rare occurrence, however, except for those caused by concomitant deficiencies in other nutrients such as zinc, iron, copper, and manganese.
POTASSIUM (K)	Yellowing between leaf veins progressing from outside of leaf toward center with occasional leaf spots; eventual scorching and curling of leaf from outside inward.	Potassium toxicity is unusual. Severe cases might cause scorching of leaf margins but would more likely be indicated by deficiencies in calcium and/or magnesium.

nitrogen deficiency, for example, while a manganese deficiency could cause brown spots like a potassium deficiency. One way to help distinguish macronutrient (NPK) deficiencies from other problems is that macronutrient deficiencies affect the older parts of the plants first, so if something is affecting only new growth, it's probably a different issue. Nitrogen deficiencies, at least, can be tested by giving more nitrogen; improvement should be rapid and dramatic. If it's not, it's probably not a nitrogen deficiency. Because nitrogen is water soluble, treating an overabundance of nitrogen might also be simple: avoid adding any more nitrogen.

Phosphorous and potassium interact with other elements in the soil and are trickier to remedy, so if you suspect they are causing problems because you've ruled out non-nutrition problems (such as insect pests or disease), bite the bullet and consider a soil test.

When and How to Use It

When and how should you fertilize? It depends on what you're growing, the conditions of your soil, and what you plan on using as fertilizer. In general, however, chemical (nonorganic) fertilizers should be applied in springtime. Organic and/or slow-acting fertilizers should be applied in the fall before planting, except for nitrogen, which will wash away; you can add that in springtime. Adding a balanced organic/slow-acting balanced fertilizer (containing nitrogen, phosphorous, and potassium) is another option. Fertilizers can be sprinkled on top of the soil, or better yet, mixed in with the top few inches of soil.

While most fertilization occurs before plants are growing, many crops will benefit from periodic or mid-season supplementation, particularly if you're using organic or otherwise slow-acting fertilizers. Two popular methods of doing so are side-dressing and foliar feeding. Side-dressing involves applying fertilizer around the base of a plant and raking it lightly into the soil. Farmers often side-dress corn with a high-nitrogen fertilizer when it's about 8 to 12 inches (20 to 30.5 cm) high, and then again when it's about 3 feet (91 cm) high. Like regular fertilization, side-dressing should not touch the plant itself.

Foliar feeding entails the direct application of dilute liquid fertilizer to a plant's foliage—the one instance when fertilizer definitely should touch the plant. Examples of foliar fertilizer include soluble synthetic products such as Miracle-Gro®, kelp extract (high in micronutrients), fish emulsion (high in nitrogen and micronutrients), and compost tea (high in beneficial microorganisms). Many farmers spray foliar fertilizer on crops every few weeks at key stages of growth (just before flowering, as first fruit forms, and so on). It is not as much a substitute for regular fertilization as it is a quick boost or a corrective of problems in the soil (such as the shortage or unavailability of micronutrients).

Beyond the Seed: Cloning Your Crops

The time will come when you will want to make more of your favorite plants to increase your crops, share with friends, sell, or for any number of other reasons. Seed-saving is an option for some plants—including many garden vegetables—but for others you'll want a clone. As much as the term may raise the specter of evil twins or galactic wars, cloning is a form of vegetative (asexual) propagation that

Seed-Saving

Seeds are the product of sexual reproduction in plants. The main advantage of sexual reproduction in plants and animals— as opposed to cloning—is that it maximizes genetic diversity. Every individual has genetic strengths and weaknesses. When plagues have struck humankind, for example, some people died and some people—often those with some advantageous genetic quirk—survived. Those are the people who lived on to reproduce and create a new generation more resistant to the plagues. And, in all likelihood, certain strains of the given plague eventually became more prevalent because they were more effective against resistant people. That's how evolution works. It's an arms race, and you want to have the biggest possible genetic arsenal.

The genetic diversity of our food crops is shrinking rapidly, however. According to the Food and Agricultural Organization of the United Nations, about 7,000 plants have been cultivated for food over the course of human history. Now about thirty crops meet 95 percent of human food needs, and nearly two-thirds of our food needs are met by just four foods: rice, wheat, corn, and potatoes. As industrial agriculture focuses on just a few cultivars, countless heirloom varieties just disappear. In fact, an estimated 75 percent of agricultural genetic diversity has disappeared since the last century. There's been an effort to prevent this erosion through massive seed collection and gene banks, but seeds are not viable forever— sometimes for just a few years, if that. The best way to preserve cultivars is to plant them and save the seeds. It also allows us to discover new cultivars.

Just as thousands of years of selective breeding transformed our ancestors' wolf enemies into man's best friend—ranging from a tiny, peach-fuzzed Chihuahua to a giant, shaggy St. Bernard—so has seed-saving produced many of our favorite fruits and vegetables. Farmers would save the seeds of the individuals with the best qualities (largest, earliest ripening, best taste, most disease resistance, and so on) and plant those. We have generations to thank for the rainbow of peppers, beans, and countless other produce we have today. Modern corn is so far removed from its original form that we can't even be sure of its parentage, and it cannot even reproduce without us.

There has long been a vibrant community of seed-savers who share forgotten and heirloom seeds with each other and even plant older varieties just to ensure that there is always a viable supply of seeds. A list of seed-saving organizations, all of which provide fantastic varieties of traditional fruits and vegetables, is included in the Notes & Resources section at the back of the book. Many of these are regional exchanges with cultivars likely to do well in their given areas, and they provide instructions on saving particular plants. The easiest vegetables to start with are from self-pollinated crops such as beans, peas, tomatoes, and peppers.

has been used for thousands of years. In fact, the theory that figs were the first domesticated plant rests, in large part, on its genetic diversity and ease of cloning. The diversity meant that some would have properties far superior to others, and the easy propagation meant that early farmers could just rip off a branch of that preferred tree and plant it somewhere else.

There are many reasons cloning might be preferred over sexual reproduction (seed-saving). Cloning can be used to produce mature individuals more quickly than through seeds and to preserve the qualities of exceptional specimens, which may be lost if you rely on sexual reproduction—think of a "bad seed." Sometimes cloning is used to propagate desired specimens of a given plant that we don't want to set seeds, as with pineapples, or that have been so modified through selective breeding that they no longer produce viable seeds, as with bananas.

Pineapples can produce viable seeds, but it's not desirable. The most common commercial varieties need to be cross-pollinated; by keeping only one cultivar around, cross-pollination—and the resulting seedy pineapple—is unlikely. Hawaii, where the fruit is a valuable crop, takes a belt-and-suspenders approach and even bans the import of pineapples' main pollinators: hummingbirds.

Unlike pineapples, commercial banana varieties no longer produce viable seeds. The forerunner of the supermarket banana had big seeds, which selective breeding has reduced to those tiny black specks you see at the base in commercial varieties. That's helpful for the eater—you won't lose a tooth biting into one—but it also means that the vestigial seed specks are not viable, and the fruit must be propagated asexually through vegetative cloning. These methods produce predictable, delicious bananas but leaves them extremely vulnerable to disease. The process of cloning can exacerbate this vulnerability; if the parent plant is infected, for example, its clonal offspring usually will be, too. As a result, we're on the verge of needing a third blockbuster variety of banana. The original commercial leader, 'Gros Michel' ('Big Mike') fell victim in the 1950s to Panama disease, caused by the same genus of fungus as fusarium wilt. 'Big Mike' was hastily replaced by the kind we find in stores today, 'Cavendish,' which was thought to be mas macho. It, too, is falling victim to Panama disease, however, and there's fear that 'Cavendish' will collapse as a commercial product in a few decades. No one quite knows which cultivar will become Banana III.

Stem/Branch Cuttings

Many methods of cloning exist, and each has its own role to play. There are three basic forms of stem or branch cuttings: softwood, semi-hardwood, and hardwood. All involve excision of a section of the plant with a growing tip, as you'd find at the end of a branch. If you're really set on propagating plants by cuttings, you can get a wealth of species-specific information from cooperative extensions and other sources. If you just want to experiment a little, following are basic instructions of each kind of cutting that will work with many plants.

Cuttings are usually about 5 inches (close to 13 cm) long with multiple nodes (the swollen areas at the tips of leaves that contain meristemic tissue). Softwood cuttings may be a bit shorter, and hardwood cuttings are often a few inches longer. The cutting is usually made just below the node of an existing leaf, which is removed. The base of the cutting—the part you want to root—is often slightly stripped of bark, incised or split using a knife, or even slightly crushed, particularly if it's a hardwood

Here, a hardwood scion is being grafted to a hardwood rootstock.

cutting. This encourages roots to grow. Rooting hormone powder is widely available and can be lightly applied to many cuttings to facilitate rooting.

Although some cuttings will root in water, most are put in a damp (not wet) sterile medium such as seed-starting mix, sand, or vermiculite. They are then covered to preserve moisture (often with a semi-inflated clear plastic bag) and placed under artificial light, in a window, or in a partially shaded spot outside. Rooting can take anywhere from days to months, with softwood cuttings rooting the fastest and hardwood cuttings the slowest.

All cuttings need to be hardened off before being put into the landscape. This involves a gradual decrease in humidity and increase in light. You might first remove the plastic bag or other covering for a day or two, and then move the plant to progressively lighter locations over a few days.

Different plants will respond better to different kinds of cutting. Those that have a reasonable chance of success by any means include figs, kiwi vines, and mulberries.

Experienced grafters can create specimens bearing several different cultivars of fruit on one tree.

Softwood Cutting

Softwood cutting is done with a plant section that has recently grown (probably spring to early summer) and has not yet hardened (in other words, it's still very flexible). Many perennial herbs, including sage, rosemary, thyme, and Vietnamese coriander, respond well to this kind of cutting. It also works well with a variety of small, bushy fruits such as blueberries, brambles, currants and gooseberries, elderberries, *Amelanchier* spp. (Juneberry), some citrus plants, and tomatoes (who knew?).

Hardwood Cutting

Hardwood fig cuttings in a plastic bottle filled with vermiculite. The clear bottle should indicate when root growth has begun, and the vermiculite permits both moisture and air flow.

Hardwood cutting is done with a plant section that is, essentially, hard wood—covered in bark and resistant to bending. Depending upon the plant, this might be a dormant section cut from the most recent season's growth or even older wood. Hardwood cuttings can be used to propagate grapes, pomegranates, and blueberries. Most stone fruits (cherries, peaches, apricots, and other tree fruit with a single pit), pomes (apples, pears, and other tree fruits with a pithy, multiseeded core), and citrus fruits are also propagated through hardwood cuttings, but not by rooting the cuttings. Instead, the cutting of the desired cultivar (known as the *scion*) is grafted onto a separately rooted plant (the rootstock or just stock). In this way, growers are able to ensure the desired characteristics of the scion, produce a fruiting tree far more quickly than could be done by seed, and sometimes use a stock with desirable characteristics other than fruiting.

Most of the grapevines in vineyards around the world, for example, have a scion of a named variety of European grape (*Vitis vinifera*)—such as 'Chardonnay' or 'Pinot Noir'—grafted onto a rootstock of American grape (such as *V. labrusca*). This is because the rootstock, while not itself producing great fruit, is very resistant to *Phylloxera*, an insect pest that devastates ungrafted *V. vinifera* vines.

Semihardwood Cuttings

Semihardwood cuttings are done with plant sections in a state somewhere between the other two, typically using the current season's growth in mid- to late summer, before it becomes dormant. Some citrus can be propagated by semi-hardwood cuttings, as can feijoa.

Root Cuttings/Division

Just as plants can be cloned from what grows above the ground, they can be cloned by what grows below. The easiest method is division, whereby you use a spade to slice off a section of a clumpy, multi-stemmed plant during dormancy and replant that section elsewhere. Division works for asparagus, rhubarb, horseradish, and most perennial herbs. A root cutting is much like one from above ground. During dormancy, you slice off a piece of healthy root several inches long and (ideally) a bit thicker than a pencil. Slash the bottom end to encourage rooting and remind you which end is which. Keep it moist at about 40 degrees Fahrenheit (4.5 degrees Celsius) (which you can do in a plastic bag filled with peat moss in your refrigerator) for a few weeks and then plant it a few inches deep in soil, either horizontally or with the intact side up. Root cuttings work for brambles, figs, horseradish, and crabapple.

Layering

Layering is the process of getting an individual plant to grow roots from an unrooted section of itself while still attached. It usually occurs in spring. The newly rooted section is then ready to be cut out and transplanted as a new individual by fall. There are several main techniques for layering:

- **Standard layering** involves burying the middle part of a flexible stem underground, with the tip above ground. If done several times with the same stem, it's called *serpentine layering*. Good for grapes.
- **Air layering** involves stripping the bark on a branch, around which you wrap a kind of bandage filled with a sterile, moist medium such as sphagnum moss. Using a wire tourniquet instead of stripping the bark is called *strangle layering*. Good for fruit trees (scions; if you want to propagate rootstocks, try mound layering).
- **Tip layer**ing involves rooting the growing tip of a stem. Good for trailing brambles.
- Mound layering involves cutting down a shrubby plant to just a few inches high, scoring the stems, and then burying about half the stems' height with soil. Good for gooseberries, figs, some rootstocks.

The wood frame behind this espalier provides both support and air flow.

Pruning and Training

Pruning and training are often used to maximize production from deciduous woody fruit plants, such as apples and stone fruits. While citrus plants don't usually require extensive training or pruning, other popular fruits do, and this offers you the chance to play Henry Higgins to an orchardful of Eliza Doolittles. There are many methods of pruning and training, but all involve ensuring good light penetration to lower branches, encouraging air circulation, removing weak or diseased growth, and promoting fruit-producing growth. These methods also aim to prevent limbs' breaking off from wind, snow cover, or the weight of fruit.

Although there are methods tailored to specific plants, there are several common to many tree crops. Freestanding trees are often pruned to the *central-leader method*, the *open-center method*, the *modified-leader method*, or *bush form*.

The central-leader method cultivates a strong central trunk beneath which ultimately grow three or four tiers of scaffold branches called *whorls*. Each whorl

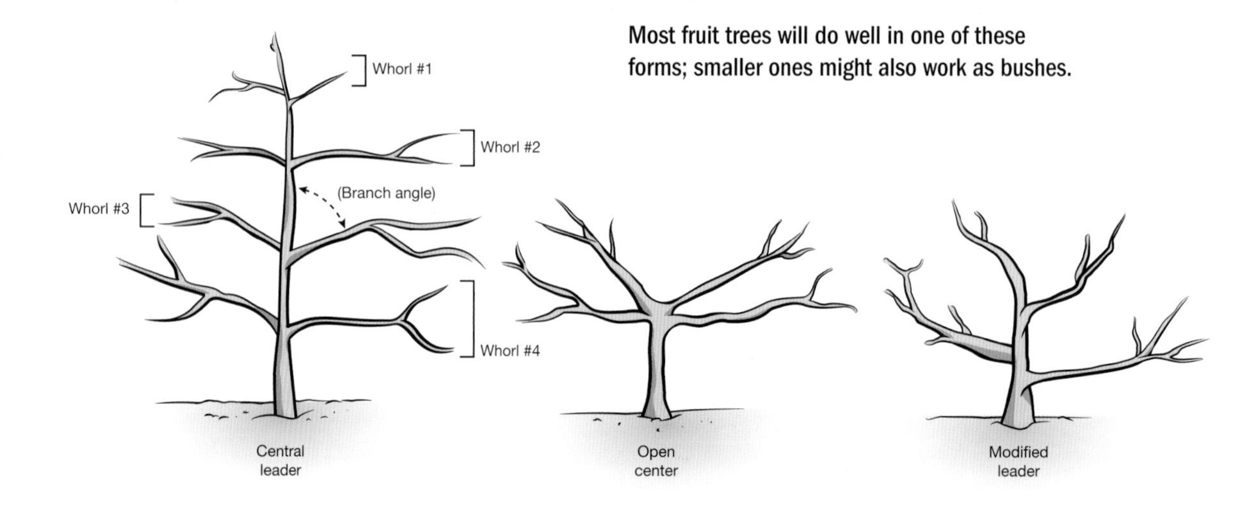

Most fruit trees will do well in one of these forms; smaller ones might also work as bushes.

consists of three or four strong branches roughly evenly spaced around the trunk (like a pinwheel or starfish). The lower the tier, the longer the branches. The central-leader approach is appropriate for trees naturally inclined to have a strong central trunk, such as apple trees. It is well suited to dwarf and semidwarf apples, resulting in a very efficient use of space.

The open-center method eliminates a central leader just a few feet off the ground, ending instead with a closely spaced whorl or two. Most stone fruits take well to open-center pruning, which results in a shorter, broader tree that is easier to harvest.

The modified-leader method takes an approach between the other two, with a central but less-dominant leader and fewer scaffold branches. Pears and persimmons adapt well to this form, as do apples and many stone fruits. It is generally considered the easiest form to master and the most widely adaptable.

Bush form is multistemmed right from the ground with an open center. It is suitable for plants that you want to keep small or that are apt to suffer winter kill (such as fig).

For the type of young trees a new urban farmer would be apt to plant, summer pruning (during active growth in spring or summer) is best, while winter pruning (in dormancy, preferably late February to early March) should be reserved mainly for correcting mistakes and removing unhealthy growth. Branch angles for most fruit trees should be in the 45–70 degree range, which is the most structurally sound. When scaffold branches grow too vertically (less than 45 degrees), you can tie them down using cords or push them down using spreaders to move them into the ideal range. When branches are not growing in quite the right direction, you can prune them back to a smaller branch or bud pointing in the right direction.

Trees do not need to be freestanding, however. Genetic dwarves will usually require some kind of support, and there are at least as many methods of training fruiting plants to grow on support systems as there are for growing freestanding plants. A basic trellis system consists of two 6- to 8-foot (almost 2- to 2.5-m) poles with a support wire running between them about 20 inches (51 cm) off the ground, and then two or three horizontal wires running evenly spaced 18 to 24 inches (46 to 61 cm) above that.

This A-frame trellis is prefect for peas, cucurbits, beans, and other viney vegetables.

Brambles are often grown between some variety of double trellis, with a set of wires running both in front of and behind the plant. Many fruits can also be espaliered—grown essentially in two dimensions against a wall, fence, or trellis. This is a good option for urban farmers looking to make use of vertical spaces, take advantage of wall microclimates, and produce attractive displays, to boot. Like well-shaped freestanding trees, supported systems require careful pruning and a solid understanding of the crops in question.

You can find information on free publications on pruning fruit trees in the Notes & Resources section; consider consulting them if you plan to grow tree crops. Practicing good pruning from the start will save you a lot of trouble, although the publications also include instructions on corrective pruning—fixing your mistakes.

Integrated Pest Management

Everyone seems to like integrated pest management (IPM), if not as a concept, at least as a sound bite. It reminds me of those "get-rich" infomercials, which always describe their products as a "system." Not a scheme, or a racket, or a complete fraud, but a "system"—suggesting something empirically proven, conceptually sound, possibly legal, almost always secret, and sure to result in a hot tub full of bikinied friends. Just throw in the word "system" and everything's okay. If only Ponzi had opted for a system rather than a scheme, we'd still be buying his book on infomercials.

So it is with IPM. It sounds good in theory ("How do you control pests in the greenhouse?" "IPM." "Oh, sure, IPM. Now there's a system."), but what does it mean in practice? As with the concept of "organic," IPM is open to vastly different interpretations. At one extreme are the people who view it essentially as an ecologically informed use of pesticides; at the other end are those for whom even the use of pesticides certified by OMRI is the equivalent of biological warfare. (This crowd sometimes derides the other crowd as practicing a different IPM: "Improved Pesticide Marketing.")

Whatever the views of people at the extremes, there's a solid core of farmers who aspire to the same thing—wielding knowledge to apply the widest variety of means to combat only target pests in the safest possible way—even if they don't agree completely on methods or even the definition of IPM. I happen to like the definition of the University of California Statewide Integrated Pest Management System, which defines it as:

> **an ecosystem-based strategy that focuses on long-term prevention of pests or their damage through a combination of techniques such as biological control, habitat manipulation, modification of cultural practices, and use of resistant varieties. Pesticides are used only after monitoring indicates they are needed according to established guidelines, and treatments are made with the goal of removing only the target organism. Pest control materials are selected and applied in a manner that minimizes risks to human health, beneficial and nontarget organisms, and the environment.**

If your concerns extend beyond merely exterminating a particular pest, you're on your way to IPM. There are tons of resources on IPM available by state, crop, and/or pest (see Notes & Resources), but at least five are basic enough to keep in mind at all times.

General Health/Cultural Practices

Nothing attracts pests as easily as a stressed or otherwise unhealthy plant. Maintaining the health of your plants is critical. Even before that, however, there is planning. Plant crops suited to your area. If you're pushing the limits of your hardiness zone—planting a fruit tree hardy in zones 4 to 7 in a zone 3 area, for example—you're inviting trouble. Yes, you can benefit from the heat-island effect and tricks such as planting along a south-facing wall, but you're still taking a bit of a gamble. Remember that for all their unsung benefits for agriculture, cities also bring well-known stressors, including air pollution and soil compaction. There's only so much one plant can take. For plants that are susceptible to certain pests generally or in your region, pick resistant cultivars or a different kind of plant. You might get lucky without doing so, but you don't want pests to establish a beachhead on your ailing peach tree and use that as a springboard to invade the rest of your garden.

Good cultural practices go hand in hand with plant health. Follow standard gardening *shoulds*. You should promptly remove and dispose of sickly plants and any fallen leaves. You should disinfect your pruning tools regularly using a 10 percent bleach solution. You should water early in the day and avoid wetting the leaves of plants vulnerable to fungal infections.

Crop rotation helps limit damage from potato beetles, which prefer nightshade crops.

Crop Diversity/Rotation

A pest in a monoculture is like a kid in a candy store—a very destructive, rapidly multiplying kid. Nature is inherently diverse, and beating out that diversity through monoculture often frees pests from their natural enemies and eliminates the pests' need for effort. Who wants to explore a wide expanse of nature for a handful of one's chosen food, when it's all provided right there, in one beautifully monotonous expanse?

Finding damage like this is heartbreaking, but identifying the pest responsible can help save your other crops.

Cultivating a wide variety of crops brings competitors and nemeses into the mix. Diversity alone might do the trick with perennial woody plants, which often cannot be moved easily, but crop rotation is paramount for herbaceous annuals. You might also purposely attract beneficial insects by keeping around some of their favorite flowering plants. Good candidates include members of the aster family (including asters, daisies, tansies, sunflowers, yarrows, and coneflowers) and umbellifers (including carrots, dill, chervil, cilantro, and Queen Anne's lace).

Monitoring

Explore your farm. Look under leaves and along stems. What's thriving and what's not? Simple as it sounds, noticing things in your garden is the first step to keeping it clean. Better than that alone is to know what you're looking for. Your local cooperative extension will have plenty of information on local pests, what they look like, and what works to combat them. This knowledge also lets you know when and where are best for finding a suspected pest. Better still is to keep records of your monitoring. Again, your local cooperative extension will likely have templates—possibly even software—to help you monitor pests. Finally, serious commercial- and

There's no substitute for hands-on surveillance. Know your enemy.

home-IPM practitioners will employ traps to accurately monitor pest pressures as a guide for treatment options. They might use Japanese beetle traps, for example, to gauge the threat of Japanese beetles by counting the caught beetles and deciding if a certain level of infestation warrants aggressive treatment.

Even if you're not that hard-core (I'm certainly not), consider what is an acceptable level of pressure from a given pest. Losing a few apples to codling moths is hardly a sacrifice if

you have scores of them, and choosing to accept that saves you some effort and the possible side effects of any more vigorous approaches.

Physical Control

Physical approaches target specific pests through essentially nonchemical means. It could mean a nighttime patrol by flashlight for snails and slugs or a high-pressure blast of water to remove aphids or mites safely. Wrapping stems or trunks in something very sticky (such as Tanglefoot®) can help keep ants and other ground-based pests away. Floating row covers do the same to air-based pests, while bird nets or deer fencing work on larger pests—at least for a while.

Chemical Means

The term chemical may evoke images of toxic waste, but let's not forget: we're made of chemicals. So are bugs. There's something very karmic about using chemicals to realize our IPM agendas. And some chemical means are, indeed, fairly innocuous. Think of slug traps that use beer as an attractant. It attracts us, too, though with

A strong jet of water is a nontoxic but effective way to deal with some pests, such as aphids.

milder effects. Synthetic pheromones use pests' own sex drives to destroy them. You can spray diluted mild soaps on some plants to make them less tasty. Other chemicals repel pests—blood meal is reported to repel deer, and kaolin-based powders repel many insects.

Then, of course, there are the killer chemicals, the "-cides." *Bacillus thuringiensis* (*Bt*), while strictly speaking a bacterium, is sort of a chemical larvicide. Sulfur and copper are often used in fungicides. "Safer" insecticides (sometimes called biopesticides) are often made of chemicals originally derived from plants, such as pyrethrins (from chrysanthemums), azadirachtin (from neem trees), or rotenone (from jicama or other sources). Other naturally derived chemicals, such as chitosan (from crustaceans), support plants' own immune systems.

Many of these are very effective and should be the vanguard of any chemical approach. Sometimes, though, you need to go nuclear. For example, Rick Bayless notes that his local fruit grower used to be organic, but it just wasn't working out. So he has turned to a conventional but restrained IPM approach. It may not be organic, but it's local and working toward sustainability.

Closely follow recommendations for all pesticides, wear protection even if the labels don't say you should, and try to make do with safer pesticides before going up the hierarchy of danger. Pesticides have a habit of being more dangerous in retrospect. The standard pesticide at the beginning of the twentieth century, for example, was arsenate of lead, a chemical containing a brilliantly deadly marriage of arsenic and lead. It reminds me of those old Reese's commercials where the folks with the peanut butter and chocolate collide, but this time in a lethal, nondelicious way. Arsenic- and lead-based inorganic pesticides were eventually supplanted by an organic one thought to be safer: DDT. Whoops!

Even rotenone, which is derived from plants, is proving to be more toxic than originally imagined. It is not specifically banned from organic agriculture, but OMRI no longer lists any products containing it. The bottom line: don't take lightly things specifically designed to kill.

Harvesting the Fruits of Your Labors

Urban farming is one of those few areas in which you can literally reap what you have sown without it meaning some sort of bitter comeuppance. Exactly how and when you harvest your crops depend on what they are, but there are some general guidelines:

Herbs: They are most potent just before flowering, and the majority can produce two or more crops per season if you cut back new growth severely (not to woody growth).

Leafy vegetables: Many leafy vegetables can be harvested as "baby" vegetables, as full-size plants, or piecemeal during the growing season. If you cut an inch or two above the soil line, the vegetable might put out new growth for a second crop.

Root vegetables: These vary. Start harvesting turnips and beetroots and their greens when roots are 1 inch (2.5 cm) in diameter. Continue until just before hot weather begins in spring for early crops and until after the first frost for late crops. Rutabagas are usually left until they reach baseball to softball size. Spring carrots

Professional growers often trellis grapes in this "candelabra" shape. It's also a popular shape to espalier fruit trees.

and parsnips should be harvested before hot weather begins, and fall ones should be harvested just before (carrots) or after (parsnips) the first freeze. Bulb onions should be kept in the ground until the tops fall over and the necks shrivel. Harvest radishes from marble size to 1 inch (2.5 cm) or more (depending on the variety), but definitely before the onset of hot weather.

Nightshades: Harvest eggplants and peppers when full size, firm, shiny, and of the desired color. Ripe tomatoes show similar characteristics, though they might have a little more give and can be ripened indoors. The husks of ripe tomatillos and ground cherries will turn brown and papery and often split open or drop when the fruit is ripe.

Cucurbits: Cucumbers and summer squashes can all be harvested when a few inches (about 7 or 8 cm) long. Larger specimens are still edible but are seedier and less desirable. Pumpkins, winter squashes, and gourds are typically harvested when the skin is hard and colorful; cut them with a knife, leaving a few inches of stem. Melons are a tricky bunch, but ripe ones of many varieties will display at least two of the following qualities: a withered or easily separated stem on one end, a softening on the other (blossom) end, and/or a faint and beguiling melon aroma.

Fruits are more inscrutable than vegetables, but here are some guidelines for a few:

Fruits to harvest ripe
(may not ripen after picking):

- berries*
- citrus, grapes
- figs*
- prickly pears*

**have short shelf lives*

Fruits that will ripen after picking:

- apples
- common pears*
- Asian pears
- persimmons*
  (Asian and American)
- kiwis*
- pawpaws*
- stone fruits

**should be picked early and brought inside to ripen*

CHAPTER 9

Urban Livestock

"Old MacDonald had a turnip" just doesn't sound right. What's a farm without animals? Many take undesirables into one end—bugs, weeds, food scraps, and the like—and send helpful manure out the other end. And, of course, one tends to eat them. Yet few of the bevy of animals on Old MacDonald's farm will cut it in the modern American city. There may be 5,000 or more cows in New Delhi, and there are hundreds of dairies in and around Dar es Salaam, Tanzania, which may provide 60 percent of that city's needs, but there's little place for Bessie in Everytown, USA. So the typical urban farmer must settle for smaller livestock and often the hazard of raising terribly cute animals for food in relatively close quarters.

It wasn't always this way. In fact, it wasn't this way until relatively recently. Some 50,000 hogs lived in Manhattan until about 1860. About 7 percent of Seattleites owned cows in 1900, including a majority in North Seattle neighborhoods. A 1906 "census of chickens" in urban areas found, on average, one clucker for every two city dwellers. In a 1932 *Los Angeles Times* article titled "There's Room for Her," the writer estimated the milk-goat population of Los Angeles County at 2,700—which would be more than 12,000 today if maintained at the same per capita level. As late as 1940, a census of

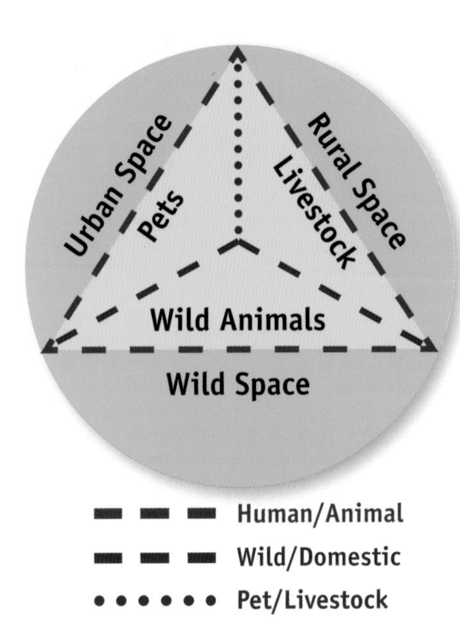

agriculture counted 352 milk cows in Brooklyn; 562 in Washington, DC; and 15,638 in Dallas County, Texas. Clearly, our urban life has changed.

Fred Brown, a historian with the National Park Service who has explored the urban livestock in Seattle for his dissertation, observes the paradox that even as cities have grown more dependent on domesticated animals—through increased meat consumption, pharmaceutical research, and other means—these animals have moved farther and farther away from our daily lives. Brown sees in the history of Seattle the interplay of the sharp Western distinctions between "human and animal, domestic and wild, pet and livestock." These "borderlands," as Brown calls them, in turn delineate the boundaries "between city and country, between home and the world of work." While we tend to see all of these distinctions as fixed and inevitable, according to Brown, they have, in fact, evolved—just like in Seattle and other cities—and remain fluid.

Closely related to the disappearance of urban agriculture in American cities has been the self-image of a "progressive," commercially minded middle class and its idea of how a city should look—aided, of course, by the advent of refrigeration and cheap transportation. Those to suffer, of course, were often the poor, immigrants, and minority groups who lived lives closer to the subsistence level. Brown notes that in the 1920s and 1930s, urban livestock were specifically excluded by restrictive covenants in developing whites-only Seattle neighborhoods, and that Federal

This is Seattle around 1905. Can you find the mostly black cow in the picture? (Hint: look up and to the right of the white telephone pole in the foreground.)

Housing Administration (FHA) loan applications from the 1930s also banned livestock on a national level.

While critics often deride contemporary urban livestock-keeping as the antithesis of its early precursors—an affectation of the elite rather than a necessity of the marginalized—one researcher has found urban livestock still popular among the marginalized for a variety of motives. In her dissertation about urban livestock—probably the most complete modern consideration of the subject in American life—Jennifer Blecha, PhD, found that contrary to many people's intuitions,

> **livestock-keepers are not primarily motivated by a desire for food. While getting fresh eggs and milk is certainly one reason for raising animals in the city, practitioners' rationales are deeper and more complex. Their livestock practices are undergirded by a dissatisfaction with dominant food and agriculture systems, a concern for the environment and animal welfare, and a desire for young people to learn compassion, care, and critical thinking.**

Keeping livestock in cities is a nuanced pursuit, and if there's one thing anathema to city ordinances, it's nuance. Your city's code may permit or forbid certain urban livestock. Or possibly both; ordinances can be contradictory. (For example, when I interviewed author and urban farmer Erik Knutzen, he pointed out that one Los Angeles ordinance permitted "truck gardening"—interpreted to permit residents to sell produce grown in their own yards—while another forbade growing crops in residential zones.)

You may be able to help change the mishmash of laws concerning urban livestock—or break them—but contend with them somehow you must. Unfortunately, there's no uniform way of finding a municipality's position on livestock, but there are some basic ways to start looking. First, see if your city's codes are online, as many are (performing a search such as "[your city] city code" or "[your city] municipal code" is a good start). Then search for terms such as "livestock," "domestic animals," "chickens," "pets," and so on. You can find some cities' restrictions regarding chickens at www.urbanchickens.org. If all else fails, call city hall. And if you get an answer you don't like, call a *different* department of city hall and see if you can get one you do like.

There are plenty of guerilla urban farmers illegally keeping livestock, but don't start keeping chickens or goats just to stick it to "the man." All due respect to the other phylogenetic kingdoms, but we're talking about fellow animals: their needs are often more complicated than plants' needs, and the neglect of them is more serious. Failure to attend to their requirements could lead to their needless suffering or death. If that doesn't motivate you, consider more practical considerations—depending on the critter and the jurisdiction, you can be fined or jailed for animal cruelty.

My family doesn't have any four-legged friends because my wife sees pets as essentially "starter children," and since we already have children, what's the point of a pet? You may not agree, but the characterization is accurate in that keeping livestock is very demanding. Are you prepared to clean out a chicken coop or milk a goat twice a day? Those chores and others await you in the world of urban livestock. You'll also need to cultivate animal-friendly friends and neighbors, or plan to never go on vacation. Take a cue from Jennie Grant (see the sidebar "A Seattle Watering Hole with a Two-Goat Minimum" later in this chapter) and consider borrowing or looking after someone else's livestock before you jump in with both feet.

Goats

It is clear why goats were one of the earliest domesticated animals and are still among the most popular. They are the Swiss Army knives of the barnyard. Like cows, goats provide meat and milk. Their meat is comparable in protein to the other big four (cow, sheep, pig, and chicken), but far lower in fat and saturated fat. Goat milk is higher in protein, minerals, and fat than cow milk and is the basis for world-class cheeses, from Greek feta to all the French chèvres. Goat hair—such as cashmere and mohair—is among the most coveted fiber in the world. Goatskin has been used for millennia, perhaps most famous in "kid gloves." Catgut, once common for sutures and the strings in stringed instruments, is actually from goats (and sheep). There are goats adapted to just about every climate, and they have even been used as pack animals. The original "horn of plenty," or cornucopia, is a goat's horn, which can also be turned into an instrument. Goats are probably smarter than any barnyard animal except, perhaps, the pig. Unlike the pig, though, goats are kosher. Should it come as any surprise that when astrologers look to the heavens, the only figure in both the Western (Capricorn) and Chinese zodiacs is a goat?

However vaunted they may be, goats are still living animals with some basic needs to be met. One idea to put out of your mind immediately is that they will gobble up tins cans and whatever spare trash you have around. Goatkeepers discover, often with a little bitterness, that goats are, if anything, a bit picky. For one thing, they favor woody plant growth (called browse) over grass or tender young greens. It may be this preference that gave them the eat-everything reputation, since they happily consume some thorny, toxic, or otherwise unsavory weeds, such as Japanese knotweed, leafy spurge, poison ivy, kudzu, thistles, and brambles. You might be able to rent them out as lawnmowers, but a better bet would be to use them to clear brush; this is a harder, presumably more profitable, job, and one better suited to their diet.

The first thing to do when considering owning a goat is to choose the kind you want and find a reputable source with healthy animals. There are hundreds of breeds of goat, commonly broken down into categories based on their best attributes: meat, milk, fiber, and so on. Milk and meat are most likely the kinds you will find in the United States, but see what varieties you can find locally. While you may have your heart set on a Stiefelgeiss or Daera Din Panah, there's a good chance that the palette of local goats will be less than universal. The National Sustainable

Jennie Grant of Seattle's Goat Justice League recommends that you if you're going to keep goats, you should keep at least two; they like company.

Agriculture Information Service lists six full-size dairy varieties as commonly available: Saanen, Alpine, Toggenburg, and Oberhasli—the so-called "Swiss breeds"—and Nubian and LaMancha.

There are two additional breeds recognized by the American Dairy Goat Association: Sable (essentially a brown Saanen) and Nigerian Dwarf. Perhaps the best option for city dwellers is a miniature dairy goat, which crosses a Nigerian Dwarf with one of the other varieties to produce a midsize model. This is the kind that Jennie Grant of Seattle's Goat Justice League keeps, and the kind now legal in Seattle because they are kind of pet-like. In addition to their helpfully small stature, mini goats produce about two-thirds as much milk as full-size ones on only half as much feed, according to the Miniature Goat Dairy Association.

Although you can eat dairy goats, too—Nubians, in particular, are meaty— other varieties are raised specifically for that purpose, including Spanish (a champion brush-clearer), Angora, Boer, and the "fainting" Tennessee Woodenlegs. The latter may be uniquely unsuited to city life because a genetic quirk renders them myotonic; they fall over in a brief state of total paralysis when startled.

Investment

A young female (foundation doe) will probably cost you at least several hundred dollars. If you want kids but not a male goat hanging around (see Challenges), you can always do what Jennie Grant does: crate up the does while they're in heat and take them to a stud (known as driveway service or the slightly more romantic pasture service). You'll need the doe to have kids if you want milk, and you should keep at least two goats anyway because goats are social.

Return

A mature milk goat can easily provide 2–4 quarts (1.9 to 3.8 liters) of milk per day for eight to ten months each year; mating is usually timed to provide the goat with a two- to four-month rest from providing milk. Angora goats can produce more than 10 pounds (4.5 kg) of mohair each year. If you're going the meat-goat route, figure that one-third of each goat's live weight will translate into marketable cuts, some with bones (rib rack) and some boneless.

Challenges

Male goats (bucks) stink, so you probably don't want them anywhere near your neighbors even if they're legal where you live. Dairy does will need to bear kids in order to lactate, so if taking your goat for occasional matchmaking offsite isn't your cup of tea, consider a different animal.

Alternatives

Goats are probably the largest vertebrates you can get away with in a modern American city; therefore, there are no good alternatives. Few individual urbanites will have the expanse of land necessary for grazing sheep, but they may be an option at the municipal level as fluffy lawnmowers, as they once were in Manhattan's Central Park.

Pot-bellied pigs, while a mere fraction of typical pig size, can still grow to much more than 150 pounds (68 kg). Milk isn't an option, and unaltered adults are reportedly temperamental, smelly, or otherwise undesirable. Pigs are also genetically quite similar to humans and susceptible to many of the same diseases, so packing them into cities probably isn't a good idea.

A Seattle Watering Hole with a Two-Goat Minimum

The first thing you'd notice talking with Jennie Grant, president of the Goat Justice League, is that she's not some crusading Legionnairess of Urban Livestock, rather just a regular person with a sense of humor and a love of animals. And it is probably precisely those qualities that helped her lead the charge to get miniature goats legalized in Seattle.

While her father had dabbled in some "bad experiments" with livestock (not realizing, for example, just how large pigs get), Grant herself had never raised livestock, let alone dreamed of infiltrating a major city with them. Besides, where could she house them on a 4,100-square-foot (381-sq-m) lot, much of which was already occupied by her bungalow? Maybe it was the fondness for animals, maybe it was a newfound taste for goat milk, but Grant eventually found her way to borrowing a friend's goats, and the rest is legislative history. Thanks to Grant's activism, which included founding the Goat Justice League, gaining some 1,000 signatures in petition, and lobbying a sympathetic city council member, § 23.42.052(D) of Seattle's Municipal Code now reads:

> Miniature Goats. The types of goats commonly known as Pygmy, Dwarf and Miniature goats may be kept as small animals, provided that male miniature goats are neutered and all miniature goats are dehorned. Nursing offspring of miniature goats licensed according to the provisions of this Code may be kept until weaned, no longer than 12 weeks from birth, without violating the limitations of subsection A.

In thanks to the city council member who helped legalize goats, Grant named the first young buckling born post-passage Richard Conlin, sadly now neutered (the goat, not the council member).

Grant now has a garden, goats, chickens, and even bees, which she decided to keep after watching a pregnant friend's hive. Grant is particularly fond of her hens, noting that "it's hard not to be a chicken collector." That's easy to believe when you see her current collection, which includes a Buff Orpington, a Red Star, and two Ameraucanas, the kind famous for laying pastel-colored eggs.

Although she is best known for her goats, Grant is also the founder of Seattle's venerable Pug Gala, devoted to the lovable, if slightly crumpled, dogs.

Rabbits

Not quite the multipurpose marvels that goats are, rabbits are nevertheless impressive, and they have an edge in terms of space requirements. Angora rabbits provide high-quality fiber, and they all provide meat, pelts, and one of the most celebrated manures, if that's not too ridiculous to say. They are even relatively low-maintenance. "In many ways," according to homesteader and author John Vivian, "rabbits are a lot less bother than chickens, and if they laid eggs, I imagine they'd take over from poultry. Rabbits don't crow or even cluck; they are perfectly happy in small cages, are easier to kill, a lot easier to dress, and ten times easier to clean up after than any kind of birds."

There are dozens of recognized rabbit breeds available, including a few different kinds of Angora and other wool-bearing rabbits. There are also several breeds that mature to 10 to 15 pounds (4.5 to 7 kg) and are favored for meat production, such as Giant Chinchillas and New Zealand rabbits. Flemish Giants can easily top 20 pounds (9 kg).

Investment

The initial cost of a rabbit can vary widely, depending upon breed and other factors. A basic setup and equipment, including a cage or hutch, nesting box, feeder, and water bottle, can cost up to $200 (or more), depending on the included features. Rabbits tolerate a wide temperature range and can usually be kept outside if given some protection in winter and shade in summer. Rabbit cages are built so that the manure falls freely, avoiding the risks posed by the rabbits' hopping around in their own waste. Underneath a rabbit cage is a preferred spot for keeping a bin of composting worms. The one time not to buy rabbits is Easter, when there is great demand and higher prices. Fertile at less than a year of age and with a roughly one-month gestation period, rabbits can reproduce very quickly. Neutering is preferable for purely pet rabbits, or if you don't need swarms of them.

Return

Rabbits have a very high "meat yield" compared with other livestock, and the meat is very lean—so lean, in fact, that death caused by overconsumption of protein without enough fat is sometimes called "rabbit starvation." Fortunately for the

squeamish, they also have more renewable uses. All rabbits produce manure that's considered among the best, and Angora and some other rabbits produce very valuable wool that you can shear off.

Challenges

Those big eyes and cuddly bodies. Could you dispatch Thumper? If so, rabbits may be your animal. I have a friend fond of serving them for Easter. If you shudder at the very notion of killing a bunny, on the other hand, and have no intention of eating hasenpfeffer, rabbits pose no other major challenges.

Alternatives

Cavies—usually known to us as guinea pigs—have been raised for meat in South America for thousands of years. They aren't truly a viable alternative in the United States, where they are thought of as pets (as with rabbits, those big eyes and cuddly bodies are quite the deterrent). But here's the argument for other countries: what they lack in volume, they make up for in numbers. They even outrabbit rabbits: a breeding pair of cavies can apparently produce more than 250 other breeding pairs within a year. In addition, cavies are generally untroubled by pests or disease.

Chickens

Chickens may be one of the best livestock choices for urban dwellers for a variety of reasons. For one thing, they don't take up a huge amount of room. They eat bugs. They immediately conjure up a rural feel (because we're no longer used to seeing them in cities). And, for the ovolactovegetarian—or just plain squeamish urban farmer looking for food without slaughter—they provide eggs. In fact, municipal codes often treat raising chickens for eggs (or as pets) more leniently than raising them for meat. Either way, chickens are one of the poster animals of the locavore movement. There are dozens of websites devoted to urban chicken devotees.

Although a typical egg-laying chicken in industrial agriculture gets about half a square foot to call her own, urban chicken enthusiasts are a more generous sort—indeed, many turn to backyard chicken keeping precisely because they dislike industrial chicken-farming practices—and recommend 2 square feet (0.2 sq m) or more of indoor space per bird for those with outside access (of at least twice that per bird) and at least 10 square feet (close to 1 sq m) per bird for confined birds (as in confined to a cage or coop, not your home).

Chickens come in an amazing and beautiful range of colors. Most breeds also have bantam versions, which are about a quarter of the standard size, and there are also some bantams without standard-size equivalents. While bantams provide less meat and smaller eggs, they also require less space and are generally thought of as hardier than their larger counterparts.

Investment

The big investment for the budding chicken farmer is the coop. It can be built from scratch, improvised from existing resources (for example, an unused shed), or bought ready for use. In fact, one brand of premade chicken house—the Eglu—is credited in part with launching the wave of urban chicken keeping in the United

An Urban Tractor

Small animals such as chickens and rabbits are often kept in something called a "chicken tractor" (or "rabbit tractor"). Imagine a long, wheeled, mesh, sometimes bottomless box that sits on the ground, and you'll have an idea of what one looks like. A chicken tractor contains a little closed-in area for the livestock and a lot of room to move around on the ground. The idea is, for example, that you wheel your chickens to a spot on the lawn and let them go to town eating weed seeds, grubs, and bugs—and dropping fertilizer—on the portion of property covered by the tractor. Then, the next day, you move the tractor to another patch. It keeps your flock happy and your lawn organically managed.

Rabbit tractors need mesh on the bottom, too, because the rabbits could burrow out. In a city, you'd also want mesh on the bottom of a chicken tractor because of what could burrow in. All enclosed animals should get water, shade, and supplemental food, if necessary. You can buy a ready-made tractor, purchase plans to build one, or just wing it yourself.

Kingdom before it caught on in the United States. The Eglu is stylish but not inexpensive. You can create your own coop for far cheaper.

Prefabricated chicken coops run from a few hundred dollars (for three or four hens) to a few thousand dollars (for fifteen or more hens). You can easily make one for much, much less, however. Feeders and waterers range in price but can also easily be homemade (just search online for "how to make" and "chicken waterer" or "chicken feeder"). What you can't make yourself are the chickens, although you may be able to get surplus

chickens from another urban chicken rancher. If you want to raise the chickens yourself, you can buy fertilized eggs, baby chicks, or young hens ready to lay (pullets).

Although chickens relish eating bugs and weed seeds, if available, you should also be prepared to feed them. Chicken feed usually comes in 50-pound (about 23-kg) bags, so be prepared to shop with a strong acquaintance if you can't lift the bag yourself.

Return

How many eggs will you get? Provided you have mature layers, you should ideally get about two eggs per day from every three chickens you have. In other words, take the number of hens you anticipate having and multiply that by 0.67 to estimate the average number of eggs per day. There's also about as many feathers and as much meat as you'd imagine. One of the nice aspects of chicken-raising is a fairly reliable market for poultry and eggs.

Challenges

The biggest challenge is probably neighbors. A hen may be no noisier than a yapping dog—probably quite a few decibels lower—but it's not a sound people are used to hearing in a city. Roosters are especially noisy and may be treated differently than hens in local zoning ordinances. In addition to human challenges, chickens and their eggs are equally loved by predators such as dogs, raccoons, and rats, so protection of the flock is paramount. The floor of the coop needs as much protection as the sides and roof because some predators will try to burrow in to get to the chickens.

Alternatives

Other domestic poultry are essentially variations on the theme of "chicken." Guinea fowl are between bantams and standard chickens in size, and they're reputedly better at eating bugs while less destructive to the garden, but they are incredibly noisy, too. Someone whose lot is crawling with bugs but is not very close to other residences might try a guinea fowl. Ducks can produce as much meat and almost as many eggs as chickens, but they are messy and helpless against predators. Geese are meatier and tougher, but they can be kind of predatory themselves (have you ever gotten "goosed"?) and a little too generous with manure. And so on and so on with turkeys, quails, and other miscellaneous poultry.

Fish and Other Aquatic Life

Fish, crustaceans, and other aquatic life have many benefits to recommend them to an urban farmer, among them the facts that they're both nutritious and delicious and provide an endless stream of free fertilizer for plant crops. Not to mention that harvesting them doesn't evoke the squeamish response that doing so with cuddly bunnies or handsome hens might. Because fish provide high-quality protein and can be raised relatively cheaply, many international agencies actively encourage "home-scale" or "small-scale" aquaculture throughout the developing world. Even that size is likely too large for most urbanites, however. A typical small-scale aquaculture pond in this scenario would be about 10 by 10 yards (about 9 x 9 m) and range from about 8 inches (20 cm) to a yard (91 cm) deep—in other words, about 15,000 gallons (almost 57,000 liters) or almost the entire volume of a small studio apartment. Based on statistics for catfish farming, you could expect about 1 pound (0.45 kg) of catfish for every 245 gallons (927 liters) of water. At the bathtub level, the most you could reasonably expect would be some edible aquatic plants and a handful of crayfish, which wouldn't make for a comfortable soak, anyway.

The real basic unit in small-scale aquaculture is a standard 55-gallon (208-liter) plastic drum or barrel. Sherilin Heise, a.k.a. Tilapia Mama, designed a system that can accommodate twelve full-size tilapia (or probably a comparable number of similarly sized fish). While you can keep an edible aquatic plant in the barrel, you need at least two, probably three, barrels to really set up a true aquaponic system. Most of these small systems recirculate the water, so much of the design involves cleaning dirty water through a combination of methods: physical filtration (through gravel or another aggregate), filtration by bacteria (which grows naturally on the gravel and other surfaces), filtration by plants (which suck up the macronutrients and other chemicals, doubling as a fertilization scheme), and oxygenation.

One very detailed online guide to building aquaculture systems of this scale is the "Barrel-Ponics" system by Travis W. Hughey. In this system only one barrel of three has fish; another has plants, and a third provides bacterial filtration or water-level management. Hughey's system is intriguing because the fish barrel is horizontal, allowing more surface area and swim room, and the plant barrel is horizontal and split in half, allowing for a decent range of crops grown in gravel. At least for tilapia, however, Heise notes that a 3-foot (91-cm) minimum depth is

Fish farming can be productive in itself as well as a profitable adjunct to vegetable farming if integrated through aquaponics.

required for them to thrive, so a shallow horizontal barrel isn't ideal. If, like Hughey, your main intention with tilapia is to fertilize plants—in an aquaponics setup or by hand—that may not be a problem.

Various species and hybrids of tilapia are usually the fish of choice in small-scale aquaculture. They tolerate small quarters and poor water quality better than many other species and are also mainly herbivorous, meaning you don't need to feed them a never-ending stream of smaller fish or pet chow. Growing Power's urban farm in Milwaukee raises tilapia and yellow perch; the former are fed miscellaneous spare vegetation and worms, and the latter commercial food and worms. On a global scale, carp species are the top choice in aquaculture; their relative unpopularity here is often attributed to pin bones that Americans don't like to pick out.

No fish or combination of fish should be chosen without some preliminary research, but this will give you a basic idea of temperature range in which a few candidate species will do relatively well: Heat-loving fish such as tilapia need to be grown inside and/or with supplemental heat in the cool winter areas, while cool-water fish such as yellow perch need protection from high summer temperatures. One advantage of a barrel system is that you can move it indoors and out again. If you plan to have the barrels outside and you pick noninsectivorous fish, you should also invest in some mosquito fish (*Gambusia* spp.) or other small insectivores to prevent your barrels from becoming neighborhood pest incubators.

APPROXIMATE TEMPERATURE RANGES FOR AQUACULTURE	65 degrees	75 degrees	85 degrees	
Yellow Perch				
Carp				
Sunfish				
Walleye				
Catfish				
Tilapia				

This gives a rough idea of water-temperature ranges tolerated by popular aquaculture fish. Note that a fish matching your temperature range doesn't mean it will match other conditions, such as water depth or stocking density. Read up on it first.

Integrating your aquacultural operation with plants in a three-or-more-barrel system is an example of aquaponics, which combines plants and animals in a mutually beneficial relationship. Growing rice and raising crayfish is a simple example of commercial aquaponics practiced (outside) in parts of the United States. Crawfish and rice enjoy similarcultural conditions, and the crawfish fertilize the rice, while the rice provides food and habitat for the crawfish. The traditional hands-down leader in aquaponics, however, is China, where farmers might combine fish, aquatic plants, ducks, mulberry trees, silkworms, and pigs into a single system that produces fish, eggs, mulberries, silk, and meat (see illustration, opposite). It's impressive, but imperfect. While aquaponics holds tremendous potential, it's yet to achieve safety, productivity, and cleanliness at the same time on a large scale. In fact, because of the complexity in creating little sustainable ecosystems, aquaponics is often a rather good refutation to the thought that one can play God. For that reason, if you're looking for a turnkey aquaculture or aquaponics system, it's probably best to buy a system from an individual or company with an extensive track record.

Investment

R&D AquaFarms offers various aquaculture and aquaponics kits for about $400 to $450, excluding shipping. If you're more of a do-it-yourselfer, however, you could

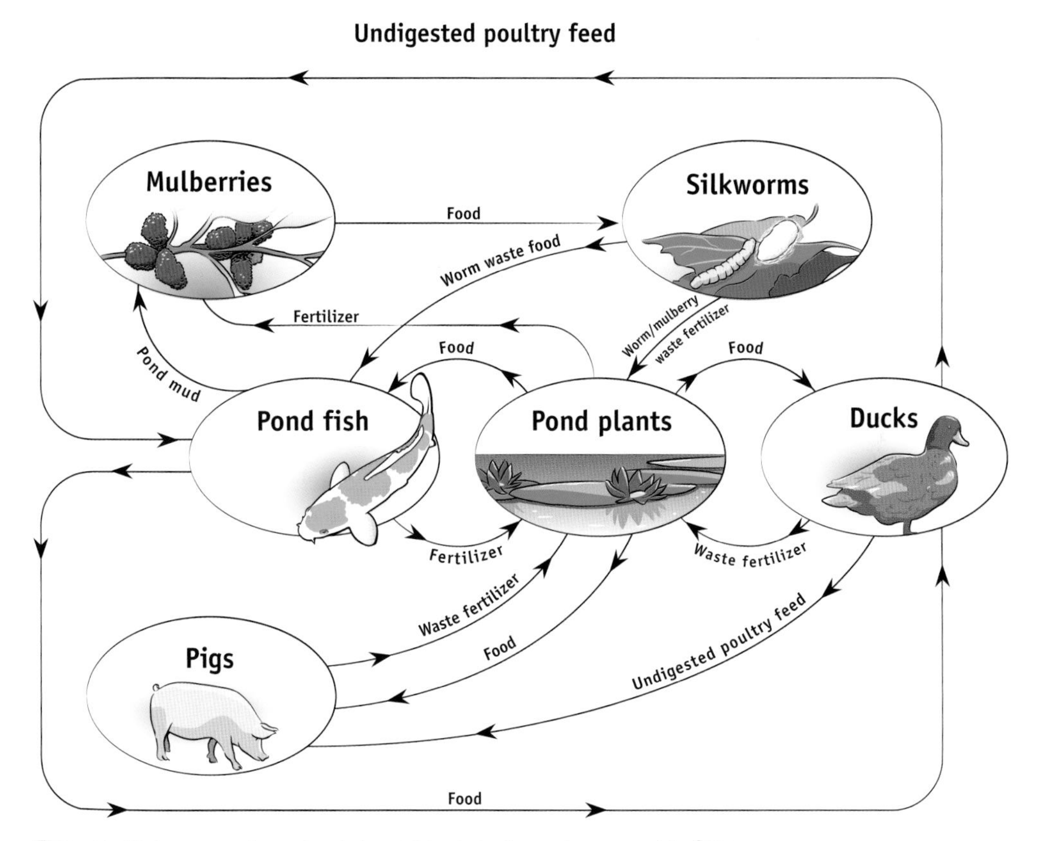

This potential aquaponics setup is based on complex systems used in China.

try building a system such as Hughey's, which would likely cost $250 to $450. The small fish you'd buy for an aquacultural or aquaponic setup are called "fingerlings" and can cost from less than a dime each to a dollar or more, depending upon species and source (see Resources). However you proceed, do some additional research first. You might start by contacting your regional aquaculture center (https://fishculture.fisheries.org/resources/regional-aquaculture-centers).

Return

The return for your investment is a steady supply of fish and top-grade (potentially) organic fertilizer for your crops. If you want fish to sell in any kind of volume rather than merely to eat, however, you'd probably need ten or more barrels, or a much larger pond or run.

Challenges

If the tenor of most primers on aquaculture could be summed up in two words, they would be "think twice." This has something to do with the technical aspects, but mainly refers to the economic ones. Can you grow enough fish to sell, and if so, would you really find buyers for tilapia á là Joe's basement? The most assured market—and according to one source, the most cost-effective—is oneself. Do you like fish?

You should consider two other factors right off the bat. First, water is heavy—50 gallons (189 liters) of water weighs more than 400 pounds (181 kg). If your only available space is an old rooftop or rickety balcony, reconsider. Second, the water temperature will need to be kept relatively constant all year long—and kept warm for many species. If you live in a region with cold winters, you'll need to either heat an outdoor tank or house your tank inside. A basement is a great location on both counts, but you would probably need artificial light.

Last, but most important, if you love fish and can make the investment and have a ridiculously sturdy place to house your setup, you still need to watch out for kids or pets. If left alone with a tank, a child can drown in minutes. Use chicken wire or whatever means necessary to make sure that's not a possibility.

Tilapia Mama

If Jennie Grant of the Goat Justice League is a champion of all creatures great and small, Sherilin Heise of B Street Growers in San Diego has somewhat of a more narrow love: tilapia. It's a focus that makes sense when you put together the elements of her life: pursuit of a PhD in science education; experience in helping set up an academic fish-research facility; gigs as extension teaching faculty at the University of California, San Diego, and as a naturalist at the San Diego Natural History Museum; and a home city just a three-hour drive away from the Salton Sea.

The Salton Sea fills a vast inland basin in southern California. It is very unusual. It is below sea level, polluted, saline, and getting saltier. Introduced fish have attracted anglers (and migratory birds) at times, but as the water quality has declined, so have many fish—but not the tilapia. Consider that the tilapia is a freshwater fish, and here it's surviving in a polluted lake that's saltier than seawater. That's tough, and this toughness and a vegetation-heavy omnivorous diet make the fish ideal for aquaculture. Heise is particularly interested in preserving the genetic diversity of these Salton Sea tilapia, which she has experience breeding, selling, and lecturing about as "Tilapia Mama."

Here are two aquaculture tips from Tilapia Mama:

1 One of the key steps in any fish setup is clearing the water of chlorine, which is especially likely to be in city systems as a part of water treatment. It kills germs but can also ravage your fish. There are many commercial products available to remove it from the water or render it harmless. Tilapia Mama suggests that there's an easier, cheaper product that you may already have in abundance: vitamin C. A 500-milligram vitamin C tablet will remove the chlorine from about a barrel of water "instantly," she says. So, as a rule of thumb, figure on about 10 milligrams per gallon (almost 4 liters).

2 Avoid snails. You'll see snails mentioned in aquaponics primers as happy cohabitants of the tank, helpfully cleaning the sides of the tank like miniature window washers. The problem is that they can be disease vectors, particularly if they are outside, where they can be exposed to birds. Land snails aren't a good option, either.

Snails

Cochlea to the Romans, *escargot* to the French, and "wall fish" to the English, the snail is good eating in any language. Heliculture, the practice of raising snails for that purpose, goes back centuries. To be an urban snail rancher in the United States, however, is strictly a hobby venture for gourmands and molluscophiles—and not just because there isn't exactly a booming commercial market for them. The handful of species generally considered the escargots all fall within the genus *Helix* (hence "heliculture"), including *H. pomatia* (apple snail or Burgundy snail) and *H. aspersa aspersa* (petit gris or little gray)—the two most popular—as well as *H. lucorum* (escargot turc or Turkish), *H. lactea* (syn. *Otala lacteal*; milk or Spanish snail), *H. aperta* (burrowing snail), and *H. aspersa maxima* (gros gris or big gray). None is native to the United States, but many can be found wild here, particularly in California and the Southeast. The only way to get them in the United States is to collect them from the wild within your state. All are banned in Canada.

One snail that is not in any legal gray area is the giant African snail, a term loosely applied to three related species: *Achatina fulica*, *A. achatina*, and *Archachatina marginata*. There's a lot to recommend them as a food source. They will eat hundreds of varieties of plants (and reportedly even stucco or paint) and grow larger than your fist, with an edible portion of 100 grams or more—essentially a quarter-pounder in a shell. But they are illegal to possess, let alone import. *Pourquoi*, you ask?

Well, even garden-variety snails can threaten important agricultural crops. African giant snails eat more than 500 plant varieties and reproduce quickly. A single hermaphroditic snail can produce more than 1,000 young in a year, each of which can produce more than 1,000 young, ad infinitum. According to the Animal and Plant Health Inspection Service (APHIS) of the US Department of Agriculture, "In 1966, a Miami, Florida, boy smuggled three giant African snails into south Florida upon returning from a trip to Hawaii. His grandmother eventually released the snails into her garden. Seven years later, more than 18,000 snails had been found along with scores of eggs." The snails were eradicated only after a decade-long, million-dollar effort.

Not surprisingly, constructing a snailery is a lot like building a prison. Snails are kept at a relatively high density (about 1 kilogram [2.2 lbs] per square meter), have a dirt yard to mill about in, have cells—such as flowerpots—to retreat to, and

Can you pick out the edible and disease-free snails from the others in a lineup? If not, then gathering wild snails is not for you.

are surrounded by a wire-mesh fence that is sometimes electrified. The Romans had to settle for surrounding them with moats.

That being said, if you love French cuisine, live in one of the few areas where you can procure legal snails, and have always imagined being a warden, snails are a good pick. They are high in protein, low in saturated fat (though high in cholesterol), and good sources of potassium, vitamin E, iron, magnesium, phosphorous, copper, and selenium. Just don't let them get away. Seriously.

The two main methods of raising snails are extensive (in a pen outside) or intensive (in a pen inside), both of which involve wire-mesh fencing to enclose the snails. This fencing should extend nearly a yard underground in outside pens to stop snails from getting out and predators from getting in. Intensive systems can have open tops if the mesh walls are curved back in to prevent escape or if you use an electric fence (6 or 12 volts through two wires a few millimeters apart; don't forget to keep the batteries fresh). Other escape-deterring options include walls constructed of or ringed by galvanized metal or copper, both of which repel snails. For intensive systems, a hard surface such as concrete suffices. Either way, snails should have a clean substrate, such as well-drained, moist soil. It should be naturally high in calcium or have supplementary calcium through bonemeal, eggshells, and so on. Snails require high humidity, and most prefer a mild and fairly constant temperature (ideally about 70 degrees Fahrenheit [21 degrees Celsius]).

There's no one-size-fits-all diet for snails, so you should try a variety of juicy fruits and vegetables—artichoke leaves are apparently a favorite among at least some species. You can try milk powder with young snails (it works with at least H. aspersa), and try to give older snails about 20 percent dry matter (bran, oats, and so on) along with the fruits and vegetables. All food should be organic. It's not a lifestyle choice for them; herbicides or pesticides can accumulate in their bodies, killing them and making them inedible. Provide fresh water in a very shallow dish (they can drown—remember the Romans with the moats?).

You can tell snails are becoming mature when, just like human teenagers, they develop lip—in the snails' case, it's around the opening in the shell.

If you're really interested in snails, you can find a lot more information on raising them—including how to prepare them for cooking—in the Notes & Resources section of this book. If you're on the fence about raising them, just be aware that the transition from snails to escargots á là bourguignonne involves copious amounts of mucus. Copious amounts.

Investment

The snails are free if you can find them. Building a snail jail can probably be done for around $100 or even less. But don't scrimp.

Return

A steady diet of below-market-rate escargots.

Challenges

Finding and identifying the snails are probably the biggest obstacles. Snails can also transmit diseases, so you should be able to identify healthy snails and know how to maintain hygienic conditions. The mucus involved in processing them is off-putting. And the risk of their escaping should never be taken lightly. Don't let them get away!

Bees

The urban beehive is very much like a city within a city. It's productive, it teems with life, and it's prepared to destroy any rube who takes it for granted. Well, honeybees are not really nasty, but realize that if you keep bees, you will get stung. Not often, perhaps, but enough to suggest that anyone with even a mild risk of anaphylaxis—or with neighbors at risk—should choose a different beast. Scranton, Pennsylvania, urban farmer Bob Philbin told me how half of his hive swarmed into a tree, so he got a ladder and scooped them down into a cardboard box. If you're not prepared for that kind of thing, reconsider. If you are, there may be a little pot of gold at the end of your rainbow.

Although there are thousands of native North American bees that can serve as effective pollinators, if you want the gold stuff, you're going to want European honeybees. There are three main "races" of honeybees—Italian, Carniolan, and Caucasian—and several hybrids. Italian honeybees are the most widely available and often considered the most docile, but they are also the least cold hardy. Check with your local cooperative extension or beekeepers' association to see which kind is best in your region. There are also three main ways to obtain honeybees: buying "package bees" from commercial sources, obtaining a nucleus colony

A bee forages on an apple-tree blossom. Bees keep agriculture buzzing—and vice versa!

(or "nuc") from another beekeeper or commercial source, or wrangling a swarm. A swarm is likely to have the most bees (read: honey potential). Package bees tend to have more members than nucs, but nucs are often better prepared to start up production and will fill out soon enough. Again, assuming that the first two are the only reasonable options for a novice, the first step you should take is to contact a local beekeeping association or beekeeper to find reputable dealers.

No matter how you obtain your bees or which kind you have, you should encircle the hive with a tall fence or hedges. Screening in the hive keeps it out of sight of potential molesters (such as kids who throw rocks at it), and also means the bees must fly up to get out—keeping their cruising altitude above the head level of people they might scare or sting. If you have a rooftop location that only you can reach, that could work, too. The bees won't mind the altitude. Another popular option is to keep bees where they are unlikely to hurt anybody: urban cemeteries. For example, honey from the 478 acres (193 hectares) in Brooklyn's historic Green-Wood Cemetery even have a cult following under the "Sweet Hereafter" label. Wherever you keep your bees, be sure to provide the with their own water source so they don't go around slurping at pools, public water fountains, and the like.

Investment

The initial investment in a hive and equipment can run up to $300, though you may be able to find used ones more cheaply. A standard 3-pound (slightly less than 1.5-kg) shipment of package bees costs up to $100; you order in fall or winter for delivery in spring. A nuc is comparable in price or slightly more expensive; it depends on what the purveyor includes with it.

Return

Honey! It's not unusual to get 20 to 40 pounds (9 to 18 kg) of it per hive every year. If you have a lot of vegetable crops needing pollination, it doesn't hurt to have a hungry army of bees around, either, especially as colony collapse disorder continues to wipe

Placing this long mason-bee nest under an overhanging sloped roof helps protect the cells from rain and wind.

out bees nationwide. Getting honey from the hive is not as simple as putting in a tap, however. Each honey-filled frame must be removed from the hive and "uncapped," which means slicing open each side with a hot knife to release the honey from the cells. Serious beekeepers use centrifuges called extractors to remove all the honey, though you can work (less efficiently) without an extractor. The remaining sheet of honeyless comb is returned to the hive, and the cappings can be melted down into beeswax, so you can expect some of that as well. (The central sheet of honey is kept intact; otherwise, the bees would need to consume several pounds of honey for each pound of wax that they need to rebuild.)

Challenges

The certainty of stings poses a risk to you, and it makes sense to minimize the risk to nosy neighbors by keeping your hive screened off. In colder areas, you'll need to take extra steps to help the bees survive the winters, such as moving the hive to a sunny spot and reducing the size of hive entrances. What do bees eat when there are no flowers around? Why, honey, of course, so make sure that you leave them some when harvesting it. You may also need to supplement with a sugar solution regularly to help honeybees tough it out.

Finally, it's a crapshoot as to whether your city permits bees or not. Plenty of urban farmers keep them irrespective of the law, but the more civic-minded approach is to petition the city to legalize them if they are currently not permitted. You can point out that legalization would put them in good company: New York City legalized beekeeping during the writing of this book, and Chicago has hives right atop city hall. Even the White House has gotten in on the action.

Alternatives

If you want the pollination benefits of bees without the honey—or the risk of stings—consider building a home for native mason bees (Osmia spp.). They live alone, survive cold winters, and rarely sting. You can easily make a nesting site for these bees by perforating an untreated block of wood with 5/16-inch (8-mm) holes about ¾ inches (almost 2 cm) apart from each other. The holes should be at least 3½ inches (9 cm) deep but not go through the back of the block; each should be a dead-end tunnel. Typical plans call for using a 6-inch (15-cm) length of a 2 by 6 (actual size 1½ x 5½ inches [38 x 140 mm])—drilling into the deep end)—or 4 by 4 (actual size 3½ x 3½ inches [89 x 89 mm]). The nest box should be at least a few feet off the ground, sheltered from wind and rain, and ideally facing east or south. The University of Maine Cooperative Extension lists several plants that enhance mason-bee populations by providing forage; fortunately for city dwellers, they include popular urban trees such as maples, oaks, dogwoods, birches, and cherries.

Nanoranching

Those with really cramped quarters (or, perhaps, a lethal allergy to bees) can manage extremely large herds of really, really small (microbial) urban livestock. Some might consider it homesteading or "adding value" rather than urban farming, but why quibble? It is the realm of culturing and fermenting, and it's diverse enough to fit either category.

Winemaking/Brewing

Winemaking is a possibility if you get bumper crops of grapes, although you can certainly make wine from other fruits and even vegetables. Elderberry, gooseberry, blueberry, cherry, and plum wines are traditional, but most any fruit or combination of fruits should do if you can balance the sugars and acids—if need be by adding sugar or acid, as they do in commercial wineries.

Beer may seem out of the question for urban farmers, since you are unlikely to harvest bushels of the barley or other cereal grains used in its manufacture. What you can harvest, however—possibly in great quantities—are hops (Humulus spp.). Hop flowers flavor and preserve beer and can go for more than five dollars per pound (even higher for organic hops), and each plant can produce a pound or more of them (enough to make a barrel or two of beer).

Hop flowers lend themselves to vertical spaces and can even be grown in large containers—which might not be bad to start with, since they are perennials that demand a sunny place with good airflow.

Fermented Dairy Products

If you raise dairy goats—or have an in with a local cattle- or goat-herder—cheesemaking is an option for the milk you can't immediately sell or consume. Jennie Grant of the Goat Justice League, for example, makes her own chèvre and mozzarella (not to mention yogurt, butter, and ice cream). Unlike regular milk production, which seems very much a volume game, cheesemaking often rewards quality and originality, offering one the chance to be "artisanal." If you can develop a reputation for your cheeses, their scarcity can increase the price that the market is willing to pay for them (ideally, at least). Since bacteria are the real workers in cheesemaking, it is not something to undertake on a lark. If you have the kind of commitment it takes to raise urban goats, however, this may not be all that challenging an adjunct.

Yogurt is not the only other alternative, either. In fact, fermenting milk predates the health-crazed 1970s by thousands of years. It does not have to be cow milk or even goat milk. Mare milk and yak milk are fermented, too, for example. And the well-known bacterium Lactobacillus acidophilus isn't the only player in town—there are other genera and species of similar lactic-acid bacteria (such as *Streptococcus thermophilus* or *Bifidobacteriim* spp.), as well as collaborating fungi (as in blue cheeses) and yeasts (as in kefir).

Store-bought yogurt contains thermophilic bacteria that thrive in warm temperatures, making a yogurt machine convenient if not strictly necessary. Easier alternatives to start with use mesophilic bacteria, which can thrive at room temperature, meaning that you don't need special equipment. Once you get a good batch going, you can add a little fresh milk to make new batches indefinitely, as long as you attend to it and no batch gets contaminated. You can get an initial bacterial dose for several different kinds of fermented, mesophilic milk cultures, including dairy kefir, buttermilk, filmjölk (Swedish), matsoni (Georgian), and piima or viili (Finnish). Although all involve some variety of lactic-acid-producing bacteria, they all taste different and have different consistencies, so you might want to try another if you're not thrilled with the first.

Fermented Nondairy Products

As with dairy products, there are many fermented nondairy ones. Two of the easiest (and perhaps most marketable) are kombucha and water kefir (also called tibicos). A kombucha starter (called a mother or scoby) looks like a pancaked mushroom but is actually a tag team of yeasts and bacteria. Water kefirs are another tag team, but they look like crystals. Both create acidic, somewhat fizzy, probiotic-rich liquids in their given media. For kombucha, it's sweetened black or green tea; for water kefir, it's various juices or flavored, sweetened liquids. Kombucha tea and water-kefir drinks can be stored easily in sealed bottles.

Unwelcome Urban Livestock

Any outdoor city farm is likely to be an attraction for, if not a home to, the less-welcome sort of urban beasts. Even if bears don't pose a hazard as they did for Bob Philbin (see chapter 4), there are plenty of other urban critters. I've heard about rabbits burrowing out sweet potatoes, raccoons devouring corn, birds decimating grapes, and squirrels methodically cleaning out sunflowers.

If you keep animals, however, you seriously raise the stakes. Birds in cages, fish in barrels, and a hiveful of honey might just be too much for a hungry urban animal to ignore. At least they're worth a shot for many animals, such as dogs and cats, skunks and raccoons, and even coyotes and larger animals. And, as one cooperative extension publication notes, addressing this by "using firearms within city limits is generally prohibited." If nothing else, you can feel a kinship with your livestock-herding ancestors, whose efforts to protect domestic animals are at least as old as agriculture. Unlike in those days, when foxes, wolves, and jackals were the most feared, the biggest culprit for urban farmers is likely to be the rat. "People need to take the rat issue more seriously," says Jennie Grant.

Fortunately, many of the steps that avoid or minimize rat problems also serve to deter other unwanted animal pests. Following are some of the ways to address the threat.

Confining Trash

Food is the number-one attraction for animal pests, and sloppy sanitation

Protecting your farm from unwanted visitors is always a consideration in cities.

A cat might eat unwanted rodents...or your chicken.

practices can keep them coming again and again. And if they're here for your leftover biscuits, why not see about making a full meal by adding your chicken? Make sure that garbage—particularly food waste—is kept in metal containers, preferably with lockable covers or fastened against stationary objects to keep them from tipping over. If you compost, make sure that no animal products (meat, oil, blood, or bone) go into the mix.

Denying Sustenance

Don't leave seeds or harvested vegetables just lying around. You should also keep an eye on plant crops; low-hanging or fallen fruit and nuts can be manna from Heaven for terrestrial animal pests. Watch out, too, for ponds or leaky hoses that might provide a watering hole for pests.

Reducing Access

Trying to close off backyard access to rats is like trying to herd cats, but fences can help stop larger predators. Slippery and/or disklike barriers on pipes, wires, and trees can at least close off some avenues of access for rats and help stop squirrels, raccoons, and the like. You can also fence off particular sections of farmstead with small-mesh hardware cloth, buried at least 6 inches deep and then another 6 inches

Try to eliminate easy access to bird feeders, which can attract all kinds of vermin.

outward underground (like a little shelf). Any livestock quarters should be very well guarded with wire mesh on all sides.

Disturbing Habitat

Rats and mice like to travel along perimeter walls and nest in loose material under shrubs and next to walls. Trim any plants providing cover near the perimeter and, if necessary, either avoid using thick mulch where rodents are problematic or (better yet) try to lay down small-mesh hardware cloth underneath a thin layer of mulch to prevent burrowing without being unsightly.

Repellents

You cannot repel all of the pests all of the time, or even all of the pests some of the time, but you can repel some of the pests some of the time. Aim for as many senses as you can: repellent tastes, repellent smells, repellent textures (such as sharp aggregate on the ground or motion-activated sprinklers), and repellent sights (such as snake or owl scarecrows and motion-activated lights). Repellent sounds used commercially (such as propane cannons) would probably be a bit extreme for the city, and you should make sure that any repellent smells are not so pungent that they offend your human neighbors. Electric fences of the sort that provide an uncomfortable shock (but not death) might be your best option in some situations, but be aware of children, roaming cats, and your city's codes.

Extreme Prejudice

Rats and mice are not only likely to be the most ubiquitous urban farming pests, but they are also most likely to be the least protected legally. Is it ever not open season on them? If you have to go this route, avoid poison baits because of the risks to children, pets, your livestock, all the food you're growing outside, and the smell of your home should a rodent choose to die there. Snap traps are probably the best option, but make sure that they are placed where children and nontarget animals won't spring them.

Chicken wire is good. Hardware cloth is better.

AFTERWORD

"Will Urban Agriculture Last?" *Fast Company* teased in 2016, "or is it just a passing, and rather unprofitable, trend?"

Fast Company's article centers on results of a survey of hundreds of American urban farmers analyzed in a 2016 study published in the *British Food Journal*. Among other things, it found that although profit was the top motive for urban farmers, only a third of them made a livelihood from urban agriculture, and the majority of farms earned less than $10,000 per year in revenue.

As the study makes clear, however, that is far from the whole story. While profit was the top reported primary motive (26 percent), that was followed by social missions regarding community (21 percent), education (10 percent), and food security (10 percent). Thus, some two-thirds of farms have primarily nonprofit motives—perhaps quite a bit more, even, since the authors excluded production of food for personal use as primary motive, as well as motives not easily classifiable.

"Given the challenges of farming in the city," the authors state, we hypothesize that the rationale for urban agriculture in the United States is only partly related to food production and farm profitability. Urban farms also appear to have nonmarket goals, including educating consumers about food and agriculture, local food systems, supporting local communities, building community, and/or reducing food insecurity in underserved neighborhoods. These social missions are closely aligned with the goals of the aforementioned food movement that is gaining traction in the United States. Thus, we posit that for many, farming in the city is a mechanism for its participants to affect change in the food system.

In fact, many farms had more than one motive; their "mission statements indicate a blurring of the profit motive and social goals, suggesting many farms are a form of social entrepreneurship." And the farms whose missions focused on improving the quality of life were more likely to put down roots in low income neighborhoods and to donate more of what they grew. They might be quite effective at achieving their social goals without being profitable.

The authors' observation certainly accords with my conversations with urban farmers. Sure, they'd like to make a living off it. But often they're happy just to be producers rather than consumers, to benefit the community, to grow their own food—or some combination of these and other reasons. And if they make a profit at the same time, all the better.

Consider two urban agriculture endeavors, Urbavore Urban farm in Kansas City, Missouri, and Hilltop Urban Farm in Pittsburgh.

Urbavore: Exit to an Urban Oasis

When we first met Brooke Salvaggio and Daniel Heryer in Chapter 6, they had been running BADSEED Farm in Kansas City, thanks in part to the good fortune that gave Brooke access to her grandfather's two and a half acres of land. With one of the biggest obstacles to urban agriculture—land tenure—being squared away, things were easy, right?

Wrong.

As BADSEED's website puts it in something of an epitaph for the farm: "In 2009, BADSEED Farm fell prey to intolerant neighbors, who attempted to shut down the farm by evoking the city's antiquated zoning ordinances. After a year-long battle, Kansas City passed an Urban Agriculture ordinance to protect urban farms. Still, BADSEED Farm was sorely compromised, confronted with the constant hostility of neighbors who prized property above community."

Even as Brooke and Daniel were headed toward their pyrrhic victory, they began searching for other large tracts of undeveloped land with Google Maps. They eventually found a 13.5 acre parcel in Kansas City's urban core whose development plans had never panned out. The city had managed it for about 20 years, with its most obvious investment being the roughly $20,000 it reportedly paid each year to mow the vacant land.

The farmers jumped through the bureaucratic hoops necessary to buy the property and found themselves with a huge new farmstead they named Urbavore, which they define as "(1) any animal that consumes urban-grown foods. (2) an organism that

feeds from the energy, surplus, and/or waste of a city."

Urbavore is huge by urban farming standards, and near the city center, but it has hardly been a cakewalk. The land came without running water or electricity, and had been ravaged by previous excavations that stripped away topsoil and left hard, compacted clay. They had to fight to get property zoned "down" to agricultural use—apparently the first time Kansas City had done that. And the land itself is a far cry from the flat, dry expanse people may picture in the Midwest: it is both sloped and incredibly wet.

Brooke and Daniel have tackled these challenges with their typical aplomb. They built a freestanding water system, for example, installed a solar array that puts far more into the city grid than they use themselves, and planted fruit trees on the slopes. Their eco-friendly home occupies one of the worst-draining spots (so as not to waste arable land), and they converted other wet fields into blueberry patches, habitat for beneficial insects, and pasture grounds for hundreds of chickens and, soon, heritage pigs.

And they have done all this while raising two sons, Percy and Solomon.

Though Brooke and Dan both began farming simply as a way to be self-sufficient, Urbavore has actually been quite lucrative. It produced $40,000 worth of food the first year, and over $500,000 in total in the years since. Still, this success has not come cheaply. "It was our sheer naivete that got us to where we are today," says Brooke. "If we had known anything about how hard this would be, we never would have tackled this project in the first place. We were 24 and stupid and ruthless. It served us well."

A Bold Rebirth in Pittsburgh

Around the same time Kansas City was parting with a virgin stretch of land to two millennial urban farmers, Pittsburgh's Housing Authority was demolishing the last of St. Clair Village, a public housing project from the 1950s. Home to nearly 1,100 families at its peak, St. Clair Village—already somewhat isolated from surrounding neighborhoods by steep terrain—had suffered depopulation and decline in recent decades. So had the surrounding St. Clair neighborhood, with high poverty and unemployment, many vacant properties, and a relatively large population of elderly residents. "St. Clair easily fits the USDA definition of a 'food desert'," according to a 2014 feasibility study by Grow Pittsburgh and Penn State Extension, "a low-income community with a poverty rate of 20 percent or more that is located one mile from a supermarket or grocery store."

Hilltop Urban Farm evolved from the idea of killing multiple birds with one stone: not just providing food but, as the study put it, "structured programming and safe green space for youth, activities for seniors, as well as the creation of entrepreneurial opportunities for Hilltop residents." In other words, an engine for community and economic development. And it has taken a community to make happen.

From the start, the project has been spearheaded by the Hilltop Alliance, an umbrella nonprofit organization comprised of community-based organizations from eleven South Pittsburgh neighborhoods, including St. Clair. The Alliance hoped to leverage the unused green space in some fashion, with urban agriculture being just one option presented to the community. In fact, one of the Alliance's first moves was to commission a "Green Toolbox Report" by Western Pennsylvania Conservancy and GTECH, which informed a four-month community engagement process and the resulting feasibility report. Local philanthropies have supported the efforts, including the Henry L. Hillman Foundation, PNC Foundation, Heinz Endowments, Birmingham Foundation, and Neighborhood Allies.

At the heart of the transformation of the St. Clair Village site will be Hilltop Urban Farm—at 23 acres, the largest in the United States. The scale of the farm aims to ensure it actually generates revenue, in addition to delivering the array of other benefits to the community.

Hilltop Urban Farm is slated to develop in phases, eventually including a youth farm, orchard, CSA fields and program, farmer incubation program, farm plots for local Bhutanese refugees and other community members, and several on-site facilities. Sixty-seven adjacent acres will remain undeveloped hillside which, along with planned stormwater mitigation ponds on the farm, will divert roughly a million gallons of water each year from flooding the sewer system. A smaller stretch of adjacent land will be saved for potential housing. Although not involved with this parallel project, Hilltop Alliance took the initiative to envision an "agri-hood" in its plans, inspired by its investigation into conservation communities such as Prairie Crossing (roughly equistant between Chicago and Milwaukee), Willowsford (in the DC metro area), and Serenbe (near Atlanta).

Hilltop Urban Farm

"I think where our project is very different than any that we've seen across the country is that the houses and community aspect would be very much an urban smart growth type of plan," says Hilltop Alliance Executive Director Aaron Sukenik. And among conservation communities, "the farm itself would also be the closest to a downtown center of any in the country." Sukenik sees this model as playing an important role as the federal government cuts back on funding to cities and urban governments assume a greater role, especially in older Rust Belt cities. "What it catalyzes is the stuff that's near impossible to finance and stimulate now, which is that urban infill.... Where I think this gets really interesting is in how you justify the potential for green space asset development like urban agriculture to leverage that longer-term reinvestment in things that grow your tax base. Because that's the path to getting more government and philanthropic funds into projects like these."

In addition to community support and some great planning efforts, Hilltop Urban Farm has another thing going for it: although it's a huge swath of Steel City, it was never an industrial site, and there's no soil contamination. St. Clair Village had been built on a farm.

*****

A passing and rather unprofitable trend? At 13.5 and 23 acres, respectively, Urbavore and Hilltop Urban Farm are hardly typical for American farms. The former is already quite profitable, and the latter shows great promise to be. Comparing them to countless smaller farms is hardly fair.

Yet both illustrate that farms can thrive without profit as the only or primary motive. Brooke Salvaggio and Daniel Heryer could have established a farm in the country more cheaply with less hassle. Yet they chose the city. Urbavore doesn't just produce food, it provides opportunities for apprentices and volunteers, and revels in participating in the local farmer's market. Urbavore contributes more clean, solar energy to the grid than the farm uses, and it offers a composting program free to metro residents that has already diverted about 1,000 tons of waste from city landfills. Hilltop Urban Farm aims to help revive a distressed neighborhood through a whole host of resources and programs.

In fact, as Carolyn Dimitri et al make clear in the *British Food Journal* study cited by Fast Company, only a minority of urban farmers are in it for the money. "For many," the authors posit, "farming in the city is a mechanism for its participants to affect change in the food system." And that certainly is a global trend, with urban agriculture increasingly seen not as a freestanding thing, but as one important plank in the wider movement of reforming food systems and reconnecting people more closely with the source of their food.

Few of us urbanites have the grit or desire to devote ourselves to urban agriculture like Brooke and Daniel, or the resources to handle the logistics of realizing an operation even a fraction of the size of Urbavore—let alone Hilltop Urban Farm. But almost all of us can join a CSA, garden a community plot, encourage our schools to incorporate more local produce, or grow a couple pots of tomatoes. And in my book, that makes us part of urban agriculture and food system reform.

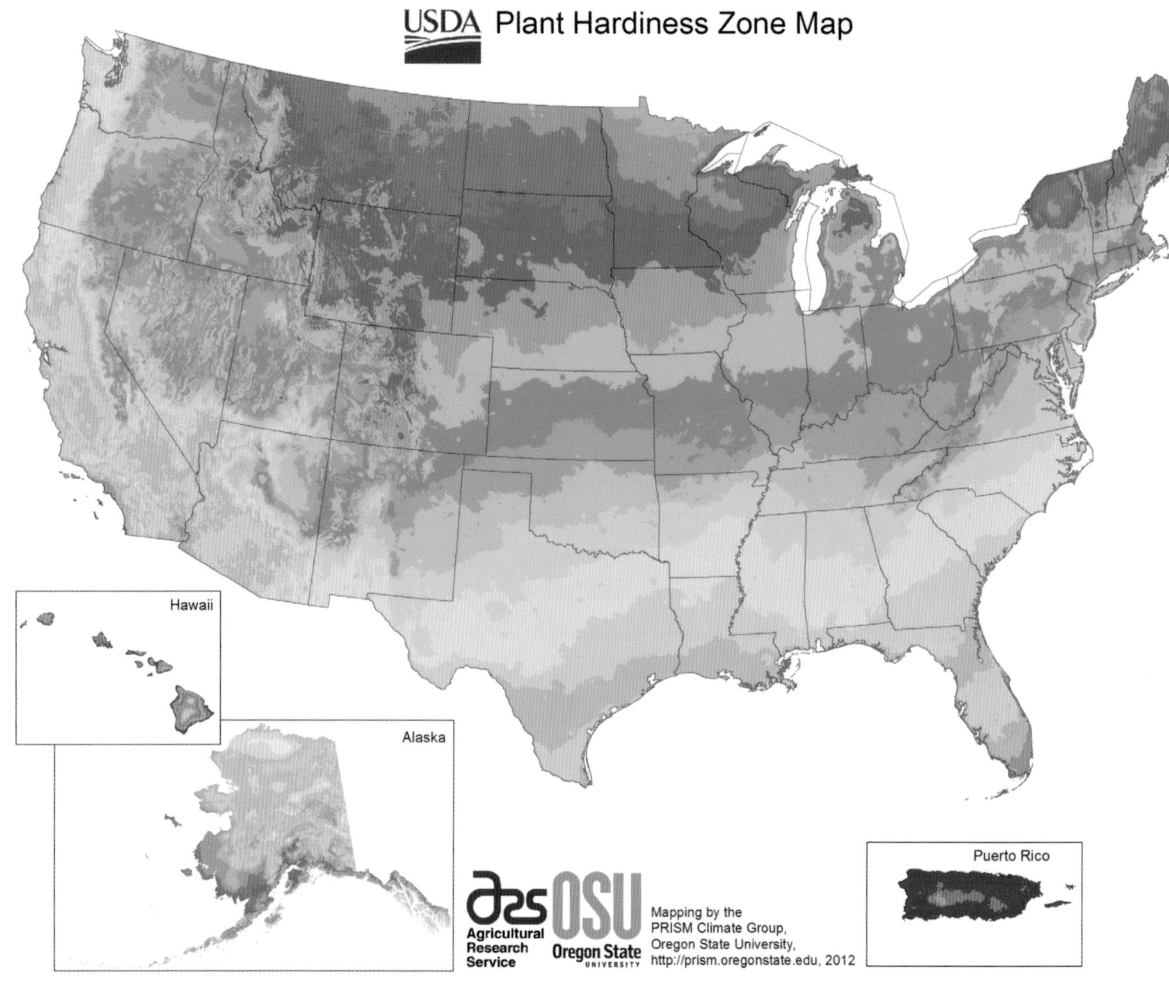

Average Annual Extreme Minimum Temperature 1976-2005

Temp (F)	Zone	Temp (C)	Temp (F)	Zone	Temp (C)
-60 to -55	1a	-51.1 to -48.3	5 to 10	7b	-15 to -12.2
-55 to -50	1b	-48.3 to -45.6	10 to 15	8a	-12.2 to -9.4
-50 to -45	2a	-45.6 to -42.8	15 to 20	8b	-9.4 to -6.7
-45 to -40	2b	-42.8 to -40	20 to 25	9a	-6.7 to -3.9
-40 to -35	3a	-40 to -37.2	25 to 30	9b	-3.9 to -1.1
-35 to -30	3b	-37.2 to -34.4	30 to 35	10a	-1.1 to 1.7
-30 to -25	4a	-34.4 to -31.7	35 to 40	10b	1.7 to 4.4
-25 to -20	4b	-31.7 to -28.9	40 to 45	11a	4.4 to 7.2
-20 to -15	5a	-28.9 to -26.1	45 to 50	11b	7.2 to 10
-15 to -10	5b	-26.1 to -23.3	50 to 55	12a	10 to 12.8
-10 to -5	6a	-23.3 to -20.6	55 to 60	12b	12.8 to 15.6
-5 to 0	6b	-20.6 to -17.8	60 to 65	13a	15.6 to 18.3
0 to 5	7a	-17.8 to -15	65 to 70	13b	18.3 to 21.1

Go to www.arborday.org to find the zone for your zip code. You can also find trees for planting in your zip code.

Canada Plant Hardiness Zone Map

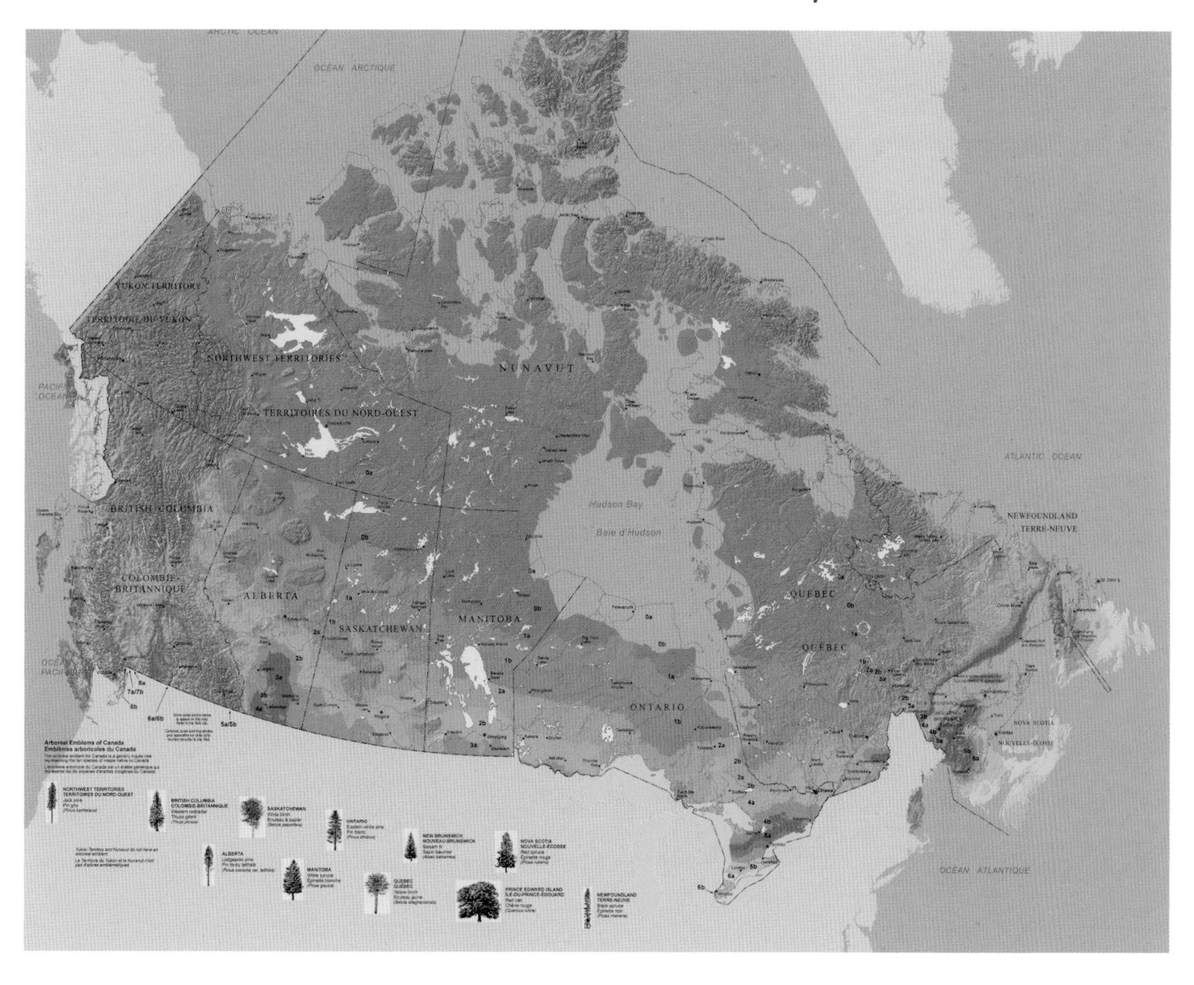

Plant hardiness zones

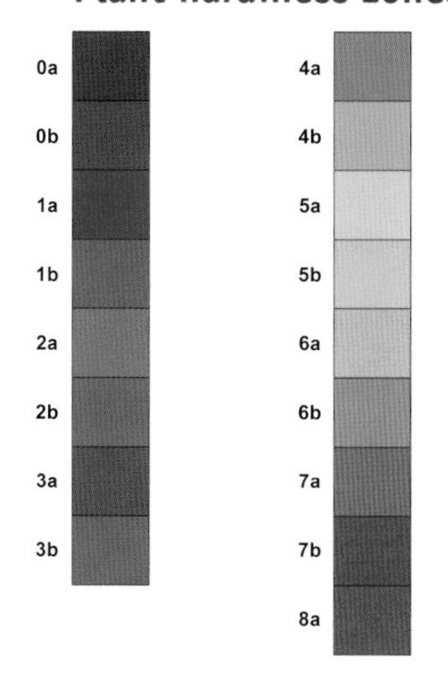

GLOSSARY

agrarian: Having to do with farming, particularly in idealizing traditional farming methods and values over contemporary industrial ones.

agritourism/agrotourism: Tourism based on agricultural attractions.

allotment: A plot of land rented out at nominal cost for someone to farm; common in European cities.

annual: A plant that completes its entire life cycle (flowering and setting seed) within a year (for example, beans, corn, and wheat).

biennial: A plant that completes its entire life cycle over two years, flowering and setting seed in the second year (for example, beets, carrots, and parsnips, although all are grown as annuals).

biodynamics: A sustainable farming method rooted in a philosophy treating the farm as an organism.

brownfield: A former industrial site, often contaminated.

commons: Land traditionally held in common; much of "the commons" became privatized in conjunction with the Industrial Revolution.

compaction: Compressive hardening of soil caused by heavy machinery, repeated foot traffic, and so on.

companion planting: Growing together two or more plants, at least one of which is beneficial to the other (more commonly, they are meant to be mutually beneficial).

compost tea: The liquid released during the composting process and/or a fertilizer solution prepared by steeping finished compost in water.

cover crop: a crop (such as alfalfa or a combination of peas and oats) used to maintain or improve soil quality.

community-supported agriculture (CSA): A program through which consumers buy "shares" of a farm's produce in advance in order to receive regular deliveries of fresh produce.

damping off: A fungal disease causing young plants to wither or an instance of its happening; often caused by overly damp soil.

espalier: To prune and train a plant to fan out in two dimensions (for example, flat against a wall) rather than three.

F_1 hybrid: A plant with uniform qualities and often hybrid vigor produced by crossing two different cultivars; subsequent generations grown from seed will not have uniform qualities.

foliar feeding: Fertilizing plants through leaves and other aboveground parts rather than through the roots.

GROW BIOINTENSIVE: A method of sustainable farming—particularly on small plots—that is taught by environmental group Ecology Action (www.growbiointensive.org) to people worldwide.

hardpan: Soil that is very difficult to work and has few pores; often caused in cities by compaction.

heirloom: A cultivar that is open-pollinated (plants grown from its seeds are just like the parents), has distinctive traits, and (often) has been passed down from generation to generation.

hilling: Adding soil regularly on top of certain crops (such as potato) during the growing season in order to encourage tuber formation.

homesteader: Someone who aims for self-sufficiency on a given property (homestead).

hybrid vigor: The superior qualities displayed by the first generation (F1) of offspring from a cross between two cultivars.

intercropping: Planting different crops in close proximity (also called multicropping or interplanting).

monoculture: Wide cultivation of a single type of crop.

new urbanism: An urban planning philosophy that prioritizes walkable cities with defined neighborhoods; somewhat opposite to urban sprawl.

open pollination: Pollination by natural means.

overwinter: To last through the cold season, particularly in regard to a plant that has been afforded some kind of protection (for example, you might overwinter a citrus plant indoors or overwinter a fig tree wrapped in burlap).

parthenocarpic: Producing fruit without fertilization, as some figs do.

perennial: A plant whose life cycle lasts more than two years (for example, apples, blueberries, and rosemary); a "tender perennial" is a perennial plant grown in a climate too cold for it to survive winter.

pollenizer: A cultivar that produces pollen effective for pollinating self-infertile cultivars of the same species.

pollinator: An animal that effects pollination by spreading pollen (usually bees, but occasionally beetles, bats, hummingbirds, or other animals).

polyculture: Cultivation of two or more crops in close proximity; often used to describe mixed crops that are somehow mutually beneficial.

row cover: A finely spun fabric used to protect crops from pests; a floating row cover is draped directly on top of plants.

shade cloth: A gauzy fabric used to protect plants from the full sun.

side-dressing: Applying (and often scratching in) fertilizer a few inches from a plant's stem; often used to supply nutrients mid-season.

soil separates: The three main components of soil—silt, clay, and sand.

SPIN-Farming: A business-oriented, franchise-like system for profitably farming small plots (www.spinfarming.com). "SPIN" stands for small plot intensive.

substrate: The medium in which plants are grown.

succession/relay planting: Planting one crop after another in the same space as another just harvested.

NOTES

In addition to what I learned from my interviews, I've relied upon a wealth of resources. To keep these notes relatively short, I have combined chapters with overlapping sources and omitted citations for facts readily confirmable by multiple sources.

Introduction and Chapters 1 & 2

7 U.S.C. §3103. "Chapter 64—Agricultural Research, Extension, and Teaching, Subchapter I—Findings, Purposes, and Definitions." www.gpo.gov

Agricultural Marketing Service. "Farmers Market Growth: 1994–2009." US Department of Agriculture. www.ams.usda.gov

Alternative Farming Systems Information Center. "Community Supported Agriculture." National Agricultural Library, US Department of Agriculture. www.nal.usda.gov

American Community Gardening Association (ACGA). "What Is a Community Garden?" www.communitygarden.org

Barstow, Cynthia. "Who's Taking Over Whom?" *The Natural Farmer*, Spring 2006. www.organicconsumers.org

Beetz, Alice. "Agroforestry Overview." National Sustainable Agriculture Information Service, 2002. http://attra.ncat.org

Berry, Wendell. "The Agrarian Standard." *Orion Magazine*, Summer 2002. www.orionmagazine.org

Blas, Javier. "US Urges Food Output Boost to Avert Unrest." FT.com, April 19, 2002.

___. "G8 Admits Losing Battle Against Hunger." FT.com, April 20, 2009.

The Biodynamic Farming and Gardening Association. "What is Biodynamic Agriculture?" www.biodynamics.com

Brown, Lester. "Could Food Shortages Bring Down Civilization?" *Scientific American*, May 2009. www.scientificamerican.com

Bruinsma, Jelle, ed. *World Agriculture: Towards 2015/2030: An FAO Perspective*. Food and Agriculture Organization of the United Nations (FAO), 2003. www.fao.org

Bruinsma, Wietse, ed., and Wilfrid Hertog, ed. *Annotated Bibliography on Urban Agriculture*. Prepared for the Swedish International Development Agency (Sida) by ETC-Urban Agriculture Programme in cooperation with TUAN and other organizations, updated March 2003. www.ruaf.org

Brundtland, Gro, chair. "Our Common Future" (the Brundtland Report). World Commission on Environment and Development, United Nations, 1987. http://worldinbalance.net

Campbell, Kurt, et al. *The Age of Consequences: The Foreign Policy and National Security Implications of Global Climate Change*. Center for Strategic & International Studies and Center for New American Security, November 2007. http://csis.org

Carlson, Allan. "Agrarianism Reborn: On the Curious Return of the Small Family Farm." *The Intercollegiate Review*, Spring 2008. www.mmisi.org

Chalk, Peter. "Replace the Weak Links in the Food Chain." *Rand Review*, Summer 2002. www.rand.org

Chez Panisse Foundation. "The Edible Schoolyard." www.chezpanissefoundation.org

Chohan, Rani. "Mississippi Dead Zone." National Aeronautics and Space Administration (NASA). www.nasa.gov

CNA Corporation. *National Security and the Threat of Climate Change*, 2007. www.securityandclimate.cna.org

Codex Alimentarius Commission. *Organically Produced Food* (third edition), *Codex Alimentarius*. Food and Agriculture Organization of the United Nations (FAO) and the World Health Organization (WHO), 2007. www.fao.org

Columbia University Mailman School of Public Health. "Dickson D. Despommier" (faculty page). www.mailman.columbia.edu

Committee on Agriculture (15th Session). *Organic Agriculture* (Item 8 of the Provisional Agenda). Food and Agriculture Organisation of the United Nations (FAO), January 25–29, 1999. www.fao.org

Danko, Danny. "The History of Hydroponics." *High Times*, May 14, 2008. http://hightimes.com

Despommier, Dickson. *The Vertical Farm Essay I* and *The Vertical Farm Essay II*. The Vertical Farm Project. www.verticalfarm.com

Diamond, Jared. *Guns, Germs, and Steel: The Fates of Human Societies*. New York: W.W. Norton & Co., 1999.

Durant, Will. *Our Oriental Heritage*. New York: Simon and Schuster, 1954.

Earles, Richard, and Paul Williams, ed. *Sustainable Agriculture: An Introduction*. National Sustainable Agriculture Information Service, 2005. http://attra.ncat.org

Ecology Action. "GROW BIOINTENSIVE: A Sustainable Solution For Growing Food," 2006. www.growbiointensive.org

Electronic Code of Federal Regulations. "Sewage sludge" (definition). Current as of April 2, 2010. www.gpo.gov

Ellis, Linden. "A China Environment Health Project Research Brief: Desertification and Environmental Health Trends in China." Woodrow Wilson International Center for Scholars, April 2, 2007. www.wilsoncenter.org

Environmental Protection Agency. "Municipal Solid Waste Generation, Recycling, and Disposal in the United States: Facts and Figures for 2006." www.epa.gov

Environmental Protection Agency. "The EPA and Food Security." Current as of May 2007. www.epa.gov

Environmental Working Group. *EWG Farm Subsidy Database Update*, 2008. http://farm.ewg.org

EurActiv Network. "Accidental GMO content permitted in organic food." June 13, 2007. www.euractiv.com

The Fertilizer Institute. www.tfi.org

Flores, H.C. *Food Not Lawns*. White River Junction, VT: Chelsea Green Publishing Company, 2006. Food and Agriculture Organization of the United Nations (FAO), "Forestry notes" in Unasylva, No. 221, Poplars and Willows, Vol. 56 2005/2. www.fao.org

Fukuoka, Masanobu, and Larry Korn, trans. *One Straw Revolution: An Introduction to Natural Farming*. New York: New York Review Books Classics, 2009.

Gershuny, Grace. "Organic at the Crossroads: Revolution or Elite Niche Market?" *The Natural Farmer*, Spring 2006. www.nofa.org

Gilman, Nils, Doug Randall, and Peter Schwartz. *Impacts of Climate Change*. GBN Global Business Network, January 2007. www.gbn.com

Glaeser, Edward. "The Lorax Was Wrong: Skyscrapers Are Green." Economix Blog, New *York Times*, March 10, 2009. http://economix.blogs.nytimes.com

Gold, Mary, and Jane Gates. *Tracing the Evolution of Organic/Sustainable Agriculture: A Selected and Annotated Bibliography*. Alternative Farming Systems.

Information Center, National Agricultural Library, United States Department of Agriculture. Updated and expanded May 2007. www.nal.usda.gov

Gurian-Sherman, Doug. *Failure to Yield: Evaluating the Performance of Genetically Engineered Crops*. Union of Concerned Scientists, April 2009. www.ucsusa.org

Harrison, Fairfax, ed. "The Husbandry of Animals," Book II of *Roman Farm Management, The Treatises of Cato and Varro, Done Into English, With Notes of Modern Instances by A Virginia Farmer*. 2nd edition, 1918. www.gutenberg.org

Hemenway, Toby. *Gaia's Garden: A Guide to Home-Scale Permaculture*, 2nd ed. White River Junction, VT: Chelsea Green Publishing Company, 2009.

Hendrickson, Mary, and William Heffernan. "2007 Concentration of Agricultural Markets Table." Food Circles Networking Project. www.foodcircles.missouri.edu

Hoffman, Leslie. *Green Roofs: Ecological Design And Construction*. Atglen, PA: Earth Pledge, Schiffer Publishing, 2004.

Holmgren, David. "Essence of Permaculture: A Summary of Permaculture Concepts and Principles Taken from Permaculture Principles & Pathways Beyond Sustainability." April 2004. www.holmgren.com.au

Horovitz, Bruce. "Recession Grows Interest in Seeds, Vegetable Gardening." *USA Today*, February 20, 2009. www.usatoday.com

Horrigan, Leo, Robert Lawrence, and Polly Walker. "How Sustainable Agriculture Can Address the Environmental and Human Health Harms of Industrial Agriculture." Environ Health Perspect 110:445–456, 2002. http://ehp03.niehs.nih.gov

Institute for Environment and Human Security. "As Ranks of 'Environmental Refugees' Swell Worldwide, Calls Grow for Better Definition, Recognition, Support." United Nations University, October 11, 2005. www.ehs.unu.edu

International Federation of Organic Agriculture Movements (IFOAM). "FAQ." www.ifoam.org International Institute of Rural Reconstruction. *Bio-intensive Approach to Small-scale Household Food Production*, 1993.

Jacke, David, and Eric Toensmeier. "Edible Forest Gardens: an Invitation to Adventure." *The Natural Farmer* (Special Supplement on AgroForestry), Spring 2002. www.nofa.org

Jacke, David. "About Edible Forest Gardens: The Ecology and Design of Home Scale Food Forests." *Edible Forest Gardens*. www.edibleforestgardens.com

Jiang, Hong. "China's Great Green Wall Proves Hollow: Tree Planting Damages Environment in Northern China." *The Epoch Times*, July 29, 2009. www.theepochtimes.com

Johnson, Nathanael. "Swine of the Times: The Making of the Modern Pig." *Harper's Magazine*, May 2006, pp. 47–56.

Koont, Sinan. "The Urban Agriculture of Havana." *Monthly Review*, January 2009. www.monthlyreview.org

Little, Jane. "The Ogallala Aquifer: Saving a Vital U.S. Water Source." *Scientific American* (special editions), March 2009. www.scientificamerican.com

LocalHarvest, Inc. "Why Buy Local?" www.localharvest.org

Locavores, "Guidelines for Eating Well." www.locavores.com

McKibben, Bill. "Where Have All the Joiners Gone?" *Orion Magazine*, March/April 2008. www.orionmagazine.org

McKinley, Jesse. "Drought Adds to Hardships in California." *New York Times*, February 21, 2009. www.nytimes.com

Mollison, Bill. Permaculture Design Course Series. *Yankee Permaculture*, 1981.

Monsanto Company. "Conservation Tillage." 2006. www.monsanto.com

Morrison, Jason, et al. *Water Scarcity & Climate Change: Growing Risks for Businesses & Investors*. Ceres and the Pacific Institute. www.ceres.org

Mougeot, Luc. "Urban Agriculture: Definition, Presence, Potentials and Risks" in *Growing Cities Growing Food*. RUAF Foundation, October 1999. www.ruaf.org

Natural Resources Conservation Council. "Conservation Practices that Save: Crop Residue Management." United States Department of Agriculture, December 2005. www.nrcs.usda.gov

Natural Resources Conservation Service. "Soil Erosion" in *National Resources Inventory* 2003 Annual NRI. United States Department of Agriculture, 2007. www.nrcs.usda.gov

New Agriculturist. "The Greening of Beijing." July 2007. www.new-ag.info

Newbury, Umut, and Megan Phelps. "Join the Real Food Revival." *Mother Earth News*, Issue 211, August/September 2005. www.motherearthnews.com

Novo, Mario, and Catherine Murphy. "Urban Agriculture in the City of Havana: a Popular Response to a Crisis" in N. Bakker, et al. *Growing Cities, Growing Food: Urban Agriculture on the Policy Agenda*. RUAF Foundation, 2001. www.ruaf.org

Organic Consumers Association. "EPA Proposes to Redefine Sewage Sludge Composts as Organic." October 4, 2005. www.organicconsumers.org

Organic Gardening. "USDA Certified Organic: What It Means To You." www.organicgardening.com

Organisation for Economic Co-Operation and Development (OECD) and Food and Agriculture

Organization of the United Nations (FAO). "Can Agriculture Meet the Growing Demand for Food?" in *OECD-FAO Agricultural Outlook 2009–2018*. 2009. www.oecdbookshop.org

Pew Center on Global Climate Change. "Agriculture Overview." 2009. www.pewclimate.org

Pew Commission on Industrial Farm Animal Production. *Putting Meat on the Table: Industrial*

Farm Animal Production in America, 2009. www.sustainabletable.org

Pew Oceans Commission. *Socioeconomic Perspectives on Marine Fisheries in the United States*,

2003. www.pewtrusts.org

Pimentel, David, and Marcia Pimentel. "Sustainability of Meat-Based and Plant-Based Diets and the Environment." *Am J Clin Nutr* 2003;78(suppl):660S–3S.

Pirog, Rich, and Andrew Benjamin. "Checking the Food Odometer: Comparing Food Miles for Local Versus Conventional Produce Sales to Iowa Institutions." Leopold Center for Sustainable Agriculture, July 2003. www.leopold.iastate.edu

Prentice, Jessica. *Full Moon Feast: Food and the Hunger for Connection*. White River Junction, VT: Chelsea Green, 2006.

___. "The Birth of Locavore." Oxford University Press blog, November 20, 2007. http://blog.oup.com

Ratliff, Evan. "The Green Wall of China." *Wired*, Issue 11.04, April 2003. www.wired.com

Riedl, Brian. "How Farm Subsidies Harm Taxpayers, Consumers, and Farmers, Too." The Heritage Foundation, June 20, 2007. www.heritage.org

Slow Food International. www.slowfood.com

Smit, Jac, Annu Ratta, and Joe Nasr. *Urban Agriculture: Food, Jobs and Sustainable Cities*. United Nations Development Program (Habitat II Series), 1996. www.energyandenvironment.undp.org

SPIN-Farming, LLC. "SPIN Frequently Asked Questions." www.spinfarming.com

The Square Foot Gardening Foundation. "How to Square Foot Garden..." www.squarefootgardening.com

Steadman, David W. "Prehistoric Extinctions of Pacific Island Birds: Biodiversity Meets Zooarchaeology," *Science*: Vol. 267, no. 5201, pp. 1123–1131, February 24, 1995. http://bio.fsu.edu

Stewart, Amy. "The Man Who Would Feed the World: John Jeavons' Farming Methods Contain Lessons For Backyard Gardeners, Too." *San Francisco Chronicle*, April 13, 2002. http://articles.sfgate.com

Sustainable Table. "Artificial hormones." www.sustainabletable.org

Union of Concerned Scientists. "Industrial Agriculture: Features and Policy." Last revised May 17, 2007. www.ucsusa.org

United Nations Population Fund (UNFPA). "The State of World Population 2001." www.unfpa.org

US Department of Agriculture. Organic Certification. www.usda.gov

US Drug Enforcement Agency. "'Hooked on Ponics' Investigation Leads to Dismantling of Marijuana Grow House Operation, Over Two Dozen Defendants Indicted." August 14, 2007. www.justice.gov

US Geological Survey. "Sources of Nutrients and Pesticides" in *The Quality of Our Nation's Waters: Nutrients and Pesticides*, US Geological Survey Circular 1225, 1999.

van Veenhuizen, René, ed. *Cities Farming for the Future: Urban Agriculture for Green and Productive Cities*. ETC-Urban Agriculture, 2006. www.idrc.ca

Vidal, John. "City Of London Plans Guerrilla Allotments for Vacant Building Sites." Guardian. co.uk, June 16, 2009.

Webber, Michael. "Energy Versus Water: Solving Both Crises Together." *Scientific American* (special editions), October 22, 2008. www.scientificamerican.com

Weber, Christopher, and H. Matthews. "Food-Miles and the Relative Climate Impacts of Food Choices in the United States." *Environ. Sci. Technol.* 2008, 42, 3508–3513. www.ncbi.nlm.nih.gov

Williams, Greg. "Permaculture: Hype or Hope?" (book review of *Gaia's Garden*). *Whole Earth*, Winter 2001.

Wu, Olivia. "Now Google's Cooking: Internet Giant's Free, Gourmet Global Cuisine Powers Its Workforce While Offering Chefs and Producers a Place to Shine." *San Francisco Chronicle*, March 1, 2006. www.sfgate.com

Yeung, Yue-man. "Examples of Urban Agriculture in Asia." *Food and Nutrition Bulletin*, The United Nations University Press, Vol. 9, No. 2, June 1987. www.unu.edu

Chapter 3

Added Value. www.added-value.org

Cai, Jianming. "Periurban Agriculture Development in China." *UA-Magazine*, Resource Centre on Urban Agriculture and Food Security (RUAF) Foundation, April 2003. www.ruaf.org

The City of Chicago's Official Tourism Site. "Green Roofs, Explore Chicago." www.explorechicago.org

Clapp, Jeni, et al. *FoodNYC: A Blueprint for a Sustainable Food System*. Office of the Manhattan Borough President, February 2010. www.libertycontrol.net

Daley, John. "Farm Initiative at Work Providing Biofuel for Local Governments." ksl.com, March 15, 2010. www.ksl.com

Earthworks Urban Farm. www.cskdetroit.org

Economic Research Service. "Foreign Agricultural Trade of the United States (FATUS)." US Department of Agriculture. www.ers.usda.gov

The Ellen Degeneres Show. "Triscuit Helps Ellen Grow a Home Farm." March 17, 2010. http://ellen.warnerbros.com

FAO's Information System on Water and Agriculture. "Haiti." Food and Agriculture Organization of the United Nations (FAO), version 2000. www.fao.org

Gallagher, John. "Is Urban Farming Detroit's Cash Cow?" *Detroit Free Press*, March 21, 2010. www.freep.com

The Garden Resource Program Collaborative. www.detroitagriculture.org

Geminder, Emily. "Greenpoint Next Frontier in Commercial-Scale Rooftop Farms." *New York Observer*, October 2, 2009. www.observer.com

General Travel China. "Modern Agricultural Science Demonstration Park at Xiaotangshan." http://english.51766.com

Goode Green. www.goodegreennyc.com

Gotham Greens. gothamgreens.com

Gray, Steven. "In Detroit, Nearly 50% Unemployment Rate?" The Detroit Blog, *Time*, December 16, 2009. http://detroit.blogs.time.com

Greensgrow Farms. www.greensgrow.org

Growing Power. www.growingpower.org

Hantz Farms. www.hantzfarmsdetroit.com

Hunts Point. www.huntspoint.com

Josar, David. "Urban Gardeners Nurture Nature in Detroit: Budding Efforts Add Green to the City's Palette." *The Detroit News*, April 24, 2009. http://detnews.com

Kansas City Center for Urban Agriculture. www.kccua.org

Kaufman, Jerry, and Martin Bailkey. *Farming Inside Cities: Entrepreneurial Urban Agriculture in the United States*. Lincoln Institute of Land Policy, 2000. www.lincolninst.edu

Kotkin, Joel. "America's Agricultural Angst" (New Geographer Column). Forbes.com, January 19, 2010. www.forbes.com

Mallach, Allan, et al. *Leaner, Greener Detroit*. American Institute of Architects Sustainable Design Assessment Team, 2008. www.aia.org

Mari Gallagher Research & Consulting Group. Examining the Impact of Food Deserts on Public Health in Detroit. 2007. www.mosesmi.org

Miller, Kathleen, and Thomas Johnson. *The Role of Agriculture and Farm Household Diversification in the Rural Economy of the United States*. Trade and Agriculture Directorate, Organisation for Economic Co-Operation and Development (OECD), July 2009. www.oecd.org

Miller, Talea. "Haiti's Farms Sow Hope for Rebuilding." *PBS Newshour*, February 12, 2010. www.pbs.org

Mukherji, Nina, and Alfonso Morales. "Zoning for Urban Agriculture." *Zoning Practice*, American Planning Association, March 2010. www.planning.org

New Orleans Food & Farm Network. www.noffn.org

Path to Freedom: The Original Modern Urban Homestead (Dervaes' Family website). http://urbanhomestead.org

Rio Grande Community Farm. www.riograndefarm.org

HAR Inc. RecoveryPark. http://recoverypark.org

Sky Vegetables, Inc. www.skyvegetables.com

Suszkiw, Jan. "Trellis-Tension Technology Fine-Tunes Grape-Yield Estimates." *Agricultural Research*, Agricultural Research Service, US Department of Agriculture, April 2006. www.ars.usda.gov

Tacio, Henrylito. "Urban Agriculture: Growing Crops in City." *Sun.Star* (Philippines), March 14, 2010. http://67.225.139.201/davao

Urban Ecological Systems Ltd. "Technology Overview." www.urbanecologicalsystems.com Urban Farm Hub. www.urbanfarmhub.org

Urban Farming. www.urbanfarming.org

Urban Harvest. "Kampala." www.uharvest.org

Valcent Products Inc. "VertiCrop High Density Vertical Growth System." www.valcent.net

van Veenhuizen, René, ed. *Cities Farming for the Future: Urban Agriculture for Green and Productive Cities*. ETC-Urban Agriculture, 2006. www.idrc.ca

Venkataraman, Bina. "Country, the City Version: Farms in the Sky Gain New Interest." *New York Times*, July 15, 2008. www.nytimes.com

Whitford, David. "Can Farming Save Detroit?" *Fortune*, December 29, 2009. http://money.cnn.com

Woyke, Elizabeth. "Material for an Architectural Revolution." *BusinessWeek*, April 24, 2007. www.businessweek.com

Yeung, Yue-man. "Examples of Urban Agriculture in Asia." *Food and Nutrition Bulletin*, The United Nations University Press, Vol. 9, No. 2, June 1987. www.unu.edu

Chapters 4 & 5

Agency for Toxic Substances & Disease Registry (ATSDR). "ToxFAQs." Centers for Disease Control and Prevention. www.atsdr.cdc.gov

American Community Gardening Association (ACGA). RebelTomato, 2007. www.communitygarden.org

AZOMITE Mineral Products, Inc. Azomite. www.azomite.com

Ball, Liz. *Composting*. New York: Smith & Hawken, Workman Publishing, 1997.

Becker, Hank. "Phytoremediation: Using Plants to Clean Up Soils." *Agricultural Research*, Agricultural Research Service, US Department of Agriculture, June 2000. www.ars.usda.gov

Berger Peat Moss. "Berger FAQ." http://bergerweb.com

Bongiorno, Lori. "Safest Plastics for Food and Beverages." *The Conscious Consumer*, Yahoo! Green, November 3, 2008. http://green.yahoo.com

Brown, Katherine, et al. *Urban Agriculture and Community Food Security in the United States: Farming from the City Center to the Urban Fringe*. Urban Agriculture Committee of the Community Food Security Coalition, October 2003. www.foodsecurity.org

Cline, Marlin, et al. *Soil Survey Manual.* Natural Resources Conservation Service, US Department of Agriculture, 1993. http://soils.usda.gov

Cornell University Cooperative Extension/ Rockland County. "Outdoor Container Gardening." January 2005. http:// rocklandcce.org

Coyne, Kelly, and Erik Knutzen. *The Urban Homestead*. Process Self-Reliance Series. Los Angeles: Process Books, 2008.

Daigle, Jean-Yves, and Hélène Gautreau-Daigle. *Canadian Peat Harvesting and the Environment* (second edition). North Ameriçan Wetlands Conservation Council Committee, Issues Paper No. 2001-1. www.peatmoss.com

Dortort, Fred. "Strange Places 6: Beach Pebbles and Moon Rocks." University of California Botanical Garden at Berkeley. http://botanicalgarden.berkeley.edu

Educational Concerns For Hunger Organization (ECHO). "An Introduction to Rooftop and Urban Gardening." www. echonet.org

Forschungsgesellschaft Landschaftsentwicklung Landschaftsbau. www.fll.de

Gershuny, Grace. *Start with the Soil.* Emmaus, PA: Rodale Press, 1993.

Gotham City Honey Co-Op. http:// gothamcitybees.com

Greentrees Hydroponics. "Rockwool." www. hydroponics.net

Hoffman, Leslie. *Green Roofs: Ecological Design and Construction*. Atglen, PA: Earth Pledge, Schiffer Publishing, 2004.

Homeyer, Henry. "On a Fad Diet of Rock Dust, How the Garden Does Grow." *New York Times*, June 24, 2004. www.nytimes.com

Hovorka, Alice, ed., Henk De Zeeuw, ed., and Mary Njenga, ed. *Women Feeding Cities: Mainstreaming Gender in Urban Agriculture and Food Security*. Practical Action, 2009. www.database.ruaf.org

Ingram, Dewayne, Richard Henley, and Thomas Yeager. *Growth Media for Container Grown Ornamental Plants*. Environmental Horticulture Department, Florida Cooperative Extension Service, Institute of Food and Agricultural Sciences, University of Florida, 2003. http://edis. ifas.ufl.edu

Jauron, Richard, and Diane Nelson. "Container Vegetable Gardening." University Extension, Iowa State University, publication PM 870B, revised May 2005. www.extension.iastate.edu

Kaufman, Jerry, and Martin Bailkey. *Farming Inside Cities: Entrepreneurial Urban Agriculture in the United States*. Lincoln Institute of Land Policy, 2000. www. lincolninst.edu

Kemper Center for Home Gardening. "Containers for Growing Plants." Missouri Botanical Garden. www.mobot.org

Kuepper, George, and Kevin Everett. *Potting Mixes for Certified Organic Production*. National Sustainable Agriculture Information Service, 2004. http://attra. ncat.org

Landis, Thomas. "Growing Media" in *The Container Tree Nursery Manual*, Volume Two: Containers and Growing Media, Forest Service, US Department of Agriculture, 1990. www.rngr.net

Long, Cheryl. "Are Old Tires Safe to Use as Planters?" *Mother Earth News*. www.motherearthnews.com

Martin, Deborah, ed., Grace Gershuny, ed., and Jerry Minnich, ed. *The Rodale Book of Composting*, rev. ed. Emmaus, PA: Rodale Press, 1992.

Matkin, O.A. *Perlite vs. Polystyrene in Potting Mixes*. The Perlite Institute, Inc. www.perlite.org McCleod, Judyth. *Botanica's Organic Gardening: The Healthy Way to Live and Grow*. Laurel Glen, 2002.

Miles, Tom. "Charcoal in Container Growing Media." *Biochar Discussion Lists and Terra Preta Website*, January 14, 2009. http://terrapreta.bioenergylists.org

Neergaard, Lauran. "Hundreds of Germs in Soil Eat Up Antibiotics." *USA Today*, April 3, 2008. www.usatoday.com

New York City Beekeepers Association. www.nyc-bees.org

NRAES (Natural Resource, Agriculture, and Engineering Service). "Characteristics of Raw Materials" (Table A.1) in *On-Farm Composting Handbook* (NRAES-54), 1992. http://compost.css.cornell.edu

Optigrün International AG. "Optigreen Green Roof Extensive One Layered Substrate Type M." www.optigreen-greenroof.com

The Perlite Institute, Inc. "Effective Watering With Horticultural Perlite." www.perlite.org

Progressive Gardening Trade Association. "Container Gardening Without Soil..." http://progressivegardening.com

Roth, Susan. *New Complete Guide to Gardening*. New York: Better Homes and Gardens Books, Meredith, 1997.

Sadowski, I.E. "Peat Bogs to Help Mitigate Climate Change." *Mother Earth News*, June/July 2001. www.motherearthnews.com

Samuel Roberts Noble Foundation. "Construction Plans for a Custom Designed Low (Mini) Tunnel." www.noble.org

The Schundler Company. "Horticultural Uses of Perlite and Vermiculite." www.schundler.com

Sea Studios Foundation. "Smart Plastics Guide." *National Geographic's Strange Days on Planet Earth*. www.pbs.org

Shayler, Hannah, Murray McBride, and Ellen Harrison. *Soil Contaminants and Best Practices for Healthy Gardens*. Cornell Waste Management Institute, October 2009. http://cwmi.css.cornell.edu

Sky Vegetables. www.skyvegetables.com

SPIN-Farming, LLC. www.spinfarming.com

Surls, Rachel, Chris Braswell, and Laura Harris. *Community Garden Start-up Guide*. University of California Cooperative Extension (UCCE)/Los Angeles County, updated March 2001 by Yvonne Savio. http://celosangeles.ucdavis.edu

Whatcom County Extension. "Cheap and Easy Worm Bin!" Washington State University.

http://whatcom.wsu.edu

Chapters 6, 7 & 8

The Cucurbit Network. Cucurbitaceae. www.cucurbit.org

Dana, Michael, and B. Rosie Lerner. "Starting Seeds Indoors." Purdue University Cooperative Extension Service, publication HO-14-W. www.hort.purdue.edu

Dave Wilson Nursery. "Apples." 2010. www.davewilson.com

___. "Fruit Varieties Developed by Zaiger's Genetics." www.davewilson.com

___. "Rootstocks." www.davewilson.com

Dean, Tamara. "Stalking the Wild Groundnut." *Orion Magazine*, November/December 2007. www.orionmagazine.org

Denham, T.P., et al. "Origins of Agriculture at Kuk Swamp in the Highlands of New Guinea." *SCIENCE*, Vol. 301, July 7, 2003. http://palaeoworks.anu.edu.au

Department of Horticulture. "Minor Fruits." Cornell University. www.hort.cornell.edu

Division of Forestry. "Persimmon (*Diospyros virginiana*)." Ohio Department of Natural Resources. www.dnr.state.oh.us

DuBose, Fred, ed. *1001 Hints & Tips for Your Garden*. Pleasantville, NY: Reader's Digest General Books, 1996.

Dufour, Rex. *Biointensive Integrated Pest Management (IPM): Fundamentals of Sustainable Agriculture*. National Sustainable Agriculture Information Service, 2001. http://attra.ncat.org

___. Biorationals: *Ecological Pest Management Database*. National Sustainable Agriculture Information Service. http://attra.ncat.org

Eames-Sheavly, Marcia, et al. *Cornell Guide to Growing Fruit at Home*. Cornell Cooperative Extension, 2003. www.gardening.cornell.edu

Earles, Richard, et al. *Organic and Low-Spray Apple Production*. National Sustainable Agriculture Information Service, October 1999. http://attra.ncat.org

Ecological Agriculture Projects. "Intercropping." McGill University, 1997. http://eap.mcgill.ca

Economic Research Service. *Fruit and Tree Nuts Outlook*. US Department of Agriculture. www.ers.usda.gov

Ells, J.E., and D. Whiting. "Saving Seed." Colorado State University Extension, publication No. 7.602, March 2008. www.ext.colostate.edu

Evans, Erv, and Jeanine Davis. "Harvesting and Preserving Herbs for the Home Gardener." North Carolina Cooperative Extension Service, publication HIL-8111, February 1998. www.ces.ncsu.edu

Fischer, David. "David Fischer's North American Mushroom Basics: Real Answers about Mushrooms (F.A.Q.)." David Fischer's American Mushrooms. http://americanmushrooms.com

Fluorescent Efficiency. "11 Ways to Save Energy and Money by Switching to LED Lighting." The Energy Superstore, July 30, 2008. www.fluorescentefficiency.com

Flynn, Paula. "Peach Leaf Curl." *Horticulture & Home Pest News*, Iowa State University, June 1991 (updated 1997 by John VanDyk). www.ipm.iastate.edu

Food and Agriculture Organization of the United Nations (FAO). "Consumption of 10 Major Vegetal Foods (2003–2005)" (table D.3) in *FAO Statistical Yearbook 2007–2008*. www.fao.org

Gershuny, Grace. *Start with the Soil*. Emmaus, PA: Rodale Press, 1993.

Golden Meadows Plant Materials Center. "Red Mulberry for Coastal Wetlands." Natural Resources Conservation Service, June 2008. www.plant-materials.nrcs.usda.gov

Gowdy, Mary, and Christopher Starbuck. "Home Propagation of Garden and Landscape Plants." University of Missouri Extension, publication G6970, May 2000. http://extension.missouri.edu

Grant, Stephanie. "A Seed in Time." *Times-News Online*, January 26, 2009. www.blueridgenow.com

Greer, Lane. *Pawpaw Production*. National Sustainable Agriculture Information Service, April

1999, Revised August 2001. http://attra.ncat.org

Guendel, Sabine, Tom Osborn, and Regina Laub. *Diversity of Experiences: Understanding Change in Crop and Seed Diversity*. Gender, Equity, and Rural Employment Division, Economic and Social Development Department, Food and Agriculture Organization of the United Nations, 2008. ftp://ftp.fao.org

Guerena, Martin, Holly Born, and Tracy Mumma. *Organic Pear Production*. National Sustainable Agriculture Information Service, April 2003. http://attra.ncat.org

Hall, M.H., and Dwight Lingenfelter. *Penn State Agronomy Guide 2009–2010*. Pennsylvania State University, CODE # AGRS-26, 2008. http://agguide.agronomy.psu.edu

Hemenway, Toby. *Gaia's Garden: A Guide to Home-Scale Permaculture*, 2nd ed. White River Junction, VT: Chelsea Green Publishing Company, 2009.

Hunter, Steven. "Re: Shitake Mushroom Cultivation." May 24, 1994. www.ibiblio.org

Janick, Jules. "The Origins of Fruits, Fruit Growing, and Fruit Breeding." *Plant Breeding Rev.*, 25:255–320 (2005). www.hort.purdue.edu

J.L. Hudson, Seedsman. "Natives Vs. Exotics: The Myth Of The Menace: Non-Native Species as Allies of Diversity." www.jlhudsonseeds.net

Kalb, Loretta. "Rancho Cordova Eyes Ordinance on Pot Growing." *Sacramento Bee*, January 10, 2010. www.sacbee.com

Kearneysville Tree Fruit Research and Education Center. "Index of Fruit Disease Photographs, Biology, and Monitoring Information." West Virginia University. www.caf.wvu.edu

Laivo, Ed. "Growing Fruit in Containers." HGTV. www.hgtv.com

Lerner, B. Rosie, and Michael Dana. "New Plants from Layering." Purdue University Cooperative Extension Service, publication HO-1-W, March 2001. www.hort.purdue.edu

Maccini, Rachel. "Seed Starting in January." University of New Hampshire Cooperative Extension. http://extension.unh.edu

Masabni, Joseph, et al. "Rootstocks for Kentucky Fruit Trees." University of Kentucky Cooperative Extension Service, revised April 2007. www.ca.uky.edu

McDorman, Bill. "Basic Seed Saving." International Seed Saving Institute, 1994. www.seedsave.org

Milius, Susan. "Landscaper's Darling Hybridizes into an Environmental Nuisance: Variation

Underlies the Callery Pear Tree's Transformation." *ScienceNews*, May 9, 2009. www.sciencenews.org

Mississippi State University Extension Service. "Growing Citrus in Containers in Mississippi." http://msucares.com

Missouri Botanical Garden. "Plants of Our Local Flora Which Have Been Used by Man as Food."

Missouri Botanical Garden Bulletin, Vol. X, No. June 6, 1922.

Mount Vernon Research Center. "Tree Fruit and Alternative Fruits for Western Washington." Washington State University. http://maritimefruit.wsu.edu

National Gardening Association. *Food Gardening Guide*. www.garden.org

National Pest Management Association. "IPM Treatment and Inspection Techniques." www.whatisipm.org

Natural Resources Conservation Service. PLANTS Database. US Department of Agriculture. http://plants.usda.gov

Neal, Rome. "Microgreens From Your Kitchen: Don't Let the Cold Stop You From Eating Fresh Vegetables." *The Early Show*, CBS, February 1, 2003. www.cbsnews.com

Nelson, Thomas, and Malcome Williamson. *Decorative Plants of Appalachia...A Source of Income*. Agriculture Information Bulletin # 342, Forest Service, US Department of Agriculture, November 1970. www.sfp.forprod.vt.edu

Niemiera, Alex. "Intensive Gardening Methods." Virginia Cooperative Extension, publication 426- 335, 2009. http://pubs.ext.vt.edu

Nordell, Anne, and Eric Nordell. "Alternative Row Covers and Cucurbits." Pennsylvania Association for Sustainable Agriculture. www.pasafarming.org

North American Fruit Explorers (NAFEX). Fruit and Nut Interest Groups. www.nafex.org

Organic Materials Review Institute (OMRI). www.omri.org

Pease, Michael. "Vegetative Erosion Barriers in Agroforestry." *The Overstory* #45. www.agroforestry.net

Peronto, Marjorie, and Theresa Guethler. "Starting Seeds at Home." University of Maine Cooperative Extension, Bulletin #2751, 2008. www.umext.maine.edu

Pesticide Management Education Program. "Crop Profile: Gooseberries in New York." Cornell Cooperative Extension, March 2000. http://pmep.cce.cornell.edu

Plant Disease Diagnostic Clinic. "Black Knot." Cornell University, April 2010. http://plantclinic.cornell.edu

Pollan, Michael. "When a Crop Becomes King." *New York Times*, July 19, 2002. www.nytimes.com

Pomper, Kirk, Sheri Crabtree, and Jeremiah Lowe. "2009 Pawpaw Cultivars and Grafted Tree

Sources." Kentucky State University. www.pawpaw.kysu.edu

Powell, M.A., T.E. Bilderback, and T.M. Disy. "Fertilizer Recommendations and Techniques to Maintain Landscapes and Protect Water Quality." North Carolina Cooperative Extension Service, Publication Number: AG-508-5, March 2006. www.bae.ncsu.edu

Ray, T. Meghan. "Japanese Flowering Apricot (*Prunus mume*)—A Mid-winter Extravaganza."

Plants & Gardens News, Vol. 10, No. 4, Winter 1995. www.bbg.org

REAP Canada. "Growing Apples Without Pesticides—Fact or Fiction?" *Alternative Agriculture News*, Vol. 9, No. 3, March 1991. http://eap.mcgill.ca

Reid, J.C. "Chef Rick Bayless: Reluctant Rock Star." *The Houston Press Food Blog*, October 7, 2009. http://blogs.houstonpress.com

Relf, Diane, Alan McDaniel, and Steve Donohue. "Fertilizing the Vegetable Garden." Virginia Cooperative Extension, publication 426–323, May 2009. http://pubs.ext.vt.edu

Relf, Diane. *Propagation by Cuttings, Layering and Division*. Virginia Cooperative Extension, publication 426–002, May 2009. http://pubs.ext.vt.edu

Santos, Bielinski, and Alicia Whidden. "Optimum Planting Dates for Intercropped Cucumber, Squash, and Muskmelon with Strawberry." Florida Cooperative Extension Service, Institute of Food and Agricultural Sciences (IFAS), University of Florida, publication HS11118, October 2007. http://edis.ifas.ufl.edu

Savio, Yvonne, and Marita Cantwell. "Prickly Pears." Washington State University Extension, Island County, 1992. www.island.wsu.edu

Savio, Yvonne. "Prickly Pear Cactus Production." Small Farm Center, University of California, revised July 1989. www.sfc.ucdavis.edu

ScienceDaily. "Potato Blight Plight Looks Promising for Food Security." August 12, 2009. www.sciencedaily.com

Scott, J.W., and B.K. Harbaugh "'Micro-Gold' Miniature Dwarf Tomato." Institute of Food and Agricultural Sciences (IFAS), University of Florida, Publication #HS987, June 2004. http://edis.ifas.ufl.edu

Sluder, Riekie. "Dye Garden Plant List." New England Unit of The Herb Society of America, updated April 2006. www.neuhsa.org

Statewide Integrated Pest Management Program. *Pests in Gardens and Landscapes—Fruit and Nuts*. Agriculture and Natural Resources, University of California. www.ipm.ucdavis.edu

Strik, B.C. "Growing Grapes in Your Home Garden." Oregon State University Extension Service, publication EC1305, reprinted June 2006. http://extension.oregonstate.edu

Terral, Jean-Frédéric, et al. "Historical Biogeography of Olive Domestication (*Olea europaea L.*) as Revealed by Geometrical Morphometry Applied to Biological and Archaeological Material." *Journal of Biogeography*, (2004) 31, 63–77. www.umr5059.univ-montp2.fr

Texas AgriLife Extension Service. "Landscape IPM." Texas A&M University. http://agrilifeextension.tamu.edu

Union of Concerned Scientists. "Close to Home: An Edible Legacy." Earthwise, Vol. 7, No. 4, Fall 2005.

University of California Division of Agriculture and Natural Resources (ANR). "Citrus Program Protects Health of State's Trees." www.ucanr.org

University of Illinois Extension. "Apple Facts." http://urbanext.illinois.edu

___. "Corn." Watch Your Garden Grow series. http://urbanext.illinois.edu

Vivian, John. *Manual of Practical Homesteading*. Emmaus, PA: Rodale Press, 1977.

The Walden Effect. "How to Cultivate Edible Mushrooms for Free." www.waldeneffect.org.

Weise, Elizabeth. "Farmers Growing Genetically Engineered Corn Break Rules." *USA Today*, November 5, 2009. www.usatoday.com

Wong, Melvin. "Visual Symptoms of Plant Nutrient Deficiencies in Nursery and Landscape Plants in Soil and Crop Management." Cooperative Extension Service, Hawaii at Manoa, publication SCM–10, January 2005. www.ctahr.hawaii.edu

Wyman, Donald. *Wyman's Gardening Encyclopedia*, expanded 2nd ed. Chicago: Scribner, 1997.

Chapter 9

Animal and Plant Health Inspection Service (APHIS). "Giant African Snail." US Department of Agriculture. www.aphis.usda.gov

Bambara, Stephen. "How To Raise and Manage Orchard Mason Bees for the Home Garden." North Carolina Cooperative Extension, publication ENT/Ort-109, revised 2002. www.ces.ncsu.edu

Blecha, Jennifer. *Urban Life with Livestock: Performing Alternative Imaginaries through Small- Scale Urban Livestock Agriculture in the United States*. University of Minnesota, July 2007.

Blue & Gray Brewing Company. "Growing Hops." www.blueandgraybrewingco.com

The Boer & Meat Goat Information Center. www.boergoats.com

The Boston Beer Company. "Samuel Adams Hop Sharing Program." www.samueladams.com

Bowman, Gail. "Meat Goats as Companion Livestock" excerpted from Bowman, Gail. *Raising Meat Goats for Profit*. Bowman Communications, 1999. www.bowmanpress.com

Brady, Emily. "Fort Wadsworth: Please Do Eat the Daisies." *New York Times*, September 20, 2007. www.nytimes.com

Brown, Frederick. "Cows in the Commons, Dogs on the Lawn: Animals, Space, and Identities in Seattle." Department of History, University of Washington, 2009. http://students.washington.edu

Bushnell, H.H. "There's Room for Her: The Goat Offers Us a Small Milk Unit That Has Earned Her a Place in Our Agricultural Economy." *Los Angeles Times*, October 13, 1932.

Center for New Crops & Plant Products. New Crop Resource Online Program. Purdue University. www.hort.purdue.edu

Clauer, Phillip. "A Small-Scale Agriculture Alternative: Poultry." Virginia Cooperative Extension, 2009. http://pubs.ext.vt.edu

Cochran, James. "Small-Scale Poultry Flocks Resources: Table of Contents/Index List." North Carolina Cooperative Extension Service, February 2006. http://robeson.ces.ncsu.edu

Coffey, Linda, Margo Hale, and Ann Wells. *Goats: Sustainable Production Overview*. National Sustainable Agriculture Information Service, August 2004. http://attra.ncat.org

Collison, Clarence, Maryann Frazier, and Dewey Caron. "Beekeeping Basics." Information and Communication Technologies in the College of Agricultural Sciences, Pennsylvania State University and Mid-Atlantic Apiculture Research and Extension Consortium (MAAREC), publication #AGRS-93, 2004. http://pubs.cas.psu.edu

Flores, H.C. *Food Not Lawns*. White River Junction, VT: Chelsea Green Publishing Company, 2006. GloryBee Foods, Inc. www.beeeducation.com

HobbyFarms.com. www.hobbyfarms.com

Hu, Bao-tong, and Hua-zhu Yang. "The Integration of Mulberry Cultivation, Sericulture and Fish Farming" (Project Report #13). Fisheries and Aquaculture Department, Food and Agriculture Organization of the United Nations (FAO), 1984. www.fao.org

Indie Hopes, LLC. "Supply Management and Pricing." www.indiehops.com

Kahn, Jeremy. "New Delhi Journal: Urban Cowboys Struggle with India's Sacred Strays." *New York Times*, November 4, 2008. www.nytimes.com

Knutzen, Erik. "Where Urban Meets Farm: Pushing City Limits." *Urban Farm Magazine*, Vol. 2, No. 1, Spring 2010.

Leeflang, Mariska. "Snail Farming in Tropical Areas." *Agromisa*, January 2005. www.agromisa.org Machen, Rick, Warren Thigpen, and Eddie Holland. "Meat Goat Carcass Fabrication for Case-Ready Products." Texas Agricultural Extension Service. http://uvalde.tamu.edu

Manci, Bill. "Prospects for Yellow Perch Aquaculture." *The Advocate*, Global Aquaculture Alliance, December 2000. http://aquanic.org

Martin, Franklin. "Guinea Pigs for Meat Production." Educational Concerns For Hunger Organization (ECHO), 1991. www.echonet.org

Miller, Laurel. "The New Chicken: Jennie Grant, the City of Seattle, and the Greatness of Goats."

The Stranger, February 2, 2010. www.thestranger.com

Miniature Dairy Goat Association. "FAQ's and Other Information about Mini-Dairy Goats." 2009. www.miniaturedairygoats.com

My Pet Chicken, LLC. www.mypetchicken.com

Nourished Kitchen. "10 Cultured Dairy Foods and How to Use Them." http://nourishedkitchen.com

R&D AquaFarms, Inc. "Helpful Tips and Information about Aquaculture and Tilapia." www.rdaquafarms.com

San Francisco Beekeepers' Association. "Beekeeping." www.sfbee.org

Siprelle, Lynn. "How to Brew Kombucha: Weird Science Project or Tasty Drink? It's Both!" *The New Homemaker*. www.thenewhomemaker.com

Smith, Nancy, and Heidi Hunt. "The Many Rewards of Rabbits: How to choose and care for these beautiful, furry creatures." *Mother Earth News*, June/July 2003. www.motherearthnews.com

Sugden, Evan, and Sylvia Kantor. "Orchard Mason Bees." Washington State University Cooperative Extension, King County, Agriculture and Natural Resources Fact Sheet #525 and Horticulture Fact Sheet #83, updated July 1999. http://king.wsu.edu

Thompson, Rebecca, and Sheldon Cheney. "Raising Snails." National Agricultural Library, US

Department of Agriculture, Special Reference Briefs Series No. SRB 96–05, July 1996. www.nalusda.gov

Tilapia Mama. "High Tech Fish Farming in Your Own Backyard: A Modern Gardening Eco-System." *Vision Magazine*, September 2009. www.visionmagazine.com

van Veenhuizen, René. *Profitability and Sustainability of Urban and Peri-Urban Agriculture*. Agricultural Management, Marketing and Finance Service, Food and Agriculture Organization of the United Nations, Occasional Paper # 19, 2007. www.fao.org

Vivian, John. *Manual of Practical Homesteading*. Emmaus, PA: Rodale Press, 1977.

Wild Blueberry Extension Office. "Field Conservation Management of Native Leafcutting and Mason Osmia Bees." University of Maine Cooperative Extension, Fact Sheet No. 301, UMaine Extension No. 2420, January 2010. www.wildblueberries.maine.edu

RESOURCES

As I've said relentlessly, the best place to look for information is your local cooperative extension. How do you find it? Gardening Know How lets you search for yours by zip code at www.gardeningknowhow.com/extension-search. (You may need to widen your search if nothing shows up within 10 miles of your location.)

Sometimes you'll want or need additional information. The following sources are among the best, places to locate that information. I've limited them to online resources to make it even easier, and you can check out my website, www.thomasjfox.com, for updates.

Agriculture Classics, Out of Print and Public Domain, Full-Text

Small Farms Library, Journey to Forever
http://journeytoforever.org/farm_library.html
Has a similar mix as the Soil and Health Library (following), with the addition of full-text books (and links to full-text works) on practical information such as building a sunflower-seed oil press or solar water heater. Arranged helpfully by topic.

Soil and Health Library
www.soilandhealth.org/01aglibrary/01principles.html
Has a mixed bag of classics and out-of-print books downloadable in PDF format. If you're looking for something from Lady Eve Balfour or Sir Albert Howard, for example, this is a good place to start.

Agriculture Topics, Brief Introductions

Beginning Farmers
http://beginningfarmers.org
A fantastic resource available through the University of Michigan that includes important but neglected topics like financing and risk management, as well as excellent and well-organized links to other resources on a whole range of issues relevant to beginning farmers.

GRIT Magazine
www.grit.com
A great resource for articles about agriculture and a good complement to *Mother Earth News*.

Hobby Farms
www.hobbyfarms.com
Hobby Farms is a great resource for hobby farmers, small production farmers, and people who just love the country. Caters to all aspects of rural life, from small farm equipment, to livestock, to crops.

Mother Earth News
www.motherearthnews.com
Jam-packed with all kinds of helpful stuff. Searching for agriculture-related terms (such as "raising chickens") is likely to produce a good number of relevant articles. I use it a lot.

Agriculture Topics, Detailed and (Often) with Further Resources

Agroforestry Net
www.agroforestry.org
Has a free bimonthly journal, *The Overstory*.

Agromisa
www.agromisa.org
Geared toward the developing world with publications in several languages.

Alternative Farming Systems Information Center (AFSIC)
https://www.nal.usda.gov/afsic
Lots of information on alternative crops and business practices, ecological pest management, etc.

Association for Temperate Agroforestry
www.aftaweb.org
A first stop for any urban farmer looking to practice agroforestry in the lower forty-eight.

Educational Concerns for Hunger Organization (ECHO)
www.echonet.org
Geared toward the developing world, but still providews plenty of good information for North America.

Faith And Sustainable Technologies (F.A.S.T.)
www.fastonline.org/CD3WD_40/CD3WD/INDEX.HTM
A gathering of hundreds of how-to manuals on topics from the mundane but useful (understanding chickens, drying fruits and vegetables) to the outlandish but fascinating (crocodile farming, making banana beer).

National Sustainable Agriculture Information Service (ATTRA)
https://attra.ncat.org
This is often my first choice for information on all things agricultural.

Sustainable Agriculture Research and Education (SARE)
www.sare.org
If you can't find anything on your sustainable agriculture topic at the National Sustainable Agriculture Information Service (ATTRA), check here.

Aquaculture/Aquaponics/Hydroponics

Barrel-Ponics (Aquaponics in a Barrel) by Travis W. Hughey, 2005
www.aces.edu/dept/fisheries/education/documents/barrel-ponics.pdf
Provides DIY instructions for making a barrel-based aquaponics system. Probably better for those geared at fertilizing crops than for those geared at producing fish to eat.

Hydroponics: An Overview by Merle H. Jensen
http://ceac.arizona.edu/pls-217
Primer on hydroponics from the University of Arizona College of Agricultural and Life Sciences.

The Northeastern Regional Aquaculture Center
https://agresearch.umd.edu/nrac/aquaculture-links
Maintains a list of other regional aquaculture centers.

An Overview of Aquaponic Systems: Hydroponic Components by D. Allen Pattillo
Published by the North Central Regional Aquaculture Center; includes extensive links to other resources.

R&D AquaFarms
www.rdaquafarms.com
Offers multiple-barrel aquaponic systems as well as small quantities of tilapia fingerlings (all male).

Climate Zone/Calendars

The National Gardening Association
https://.garden.org/zipzone
Enables you to search for your climate zone by zip code.
https://.garden.org/apps/calendar
Generates a printable planting calendar for popular fruits and vegetables according to your city or zip code.

The Old Farmer's Almanac
www.almanac.com/gardening/planting-dates
Generates a printable planting calendar for popular fruits and vegetables according to your city or zip code. You can also sign up for email reminders.

Community Gardens, Kitchen Gardens, CSAs, and Farmers Markets

The Agricultural Marketing Service of the US Department of Agriculture
http://apps.ams.usda.gov/FarmersMarkets
The site has a searchable database of farmers' markets.

American Community Gardening Association (ACGA)
www.communitygarden.org
All you ever wanted to know about community gardens, including a searchable (but not complete) database and links to start-up resources, including the Community Garden Start-Up Guide put out by the University of California Cooperative Extension (UCCE) in Los Angeles.

Local Harvest
www.localharvest.org
Provides a searchable database of farms and CSAs, information on events, and other resources.

The Rodale Institute
www.rodaleinstitute.org/farm_locator
Provides a searchable database of farmers.

Cultivars

Cornell Cooperative Extension
http://vegvariety.cce.cornell.edu/help.php
Offers a great searchable database of vegetable cultivars.

Kentucky State University
www.pawpaw.kysu.edu/pawpaw/cvsrc98.htm
Maintains a list of pawpaw cultivars and grafted-tree sources.

Distance Learning

Distance Learning on Urban Agriculture
http://moodle.ruaf.org
RUAF, ETC-Urban Agriculture (an advisory group and resource center based in the Netherlands), and the Ryerson University's G. Raymond Chang School of Continuing Education and Centre for Studies in Food Security have developed this distance-learning program in urban agriculture. Users can take the courses at their own paces for free (although without credit or certificate), or they can take them as paid and accredited courses.

Edible Schoolyards, Etc.

The Edible Schoolyard Project
www.edibleschoolyard.org
The site of the original Edible Schoolyard, with links to resources to help parents and teachers.

National Farm to School Network
www.farmtoschool.org
A national program linking farms with their local schools to nourish children, strengthen farms, and otherwise improve communities.

National Gardening Association's KidsGardening
www.kidsgardening.org
A child-focused initiative of the National Gardening Association.

Food Policy Councils and Other Food Issues

Food Tank
https://foodtank.com/
"The think tank for food."

The Institute for Food and Development Policy (Food First)
www.foodfirst.org
Provides information on various food issues.

Slow Food USA
www.slowfoodusa.org
Provides information on food and links to local chapters.

The Union of Concerned Scientists
www.ucsusa.org/food_and_agriculture
Provides science-based information on food and agriculture.

Gardening Equipment and Supplies

EarthBox
www.earthbox.com
A great way to start growing your garden right away is to use an EarthBox system.

Lee Valley & Veritas Tools
www.leevalley.com
This is my favorite source for all kinds of high-quality tools and other gardening supplies. They have all kinds of items hard to find elsewhere, such as hinges for raised beds, grow pots that fit right into bags of soil, and brick-gripping clips that allow you to hang things from brick walls.

Green Roofs

Green Roofs for Healthy Cities (GRHC)
www.greenroofs.org
The national industry association for green roofs in the United States; a good place to start if you're thinking of building one.

International Green Roof Association
www.igra-world.com
The international counterpart of GRHC.

Hoop-House Instructions

The Samuel Roberts Noble Foundation
https://www.noble.org/globalassets/docs/ag/pubs/horticulture/nf-ho-12-02.pdf
Provides instructions on building a 14-foot-high tunnel up to 80 feet long.

Utah State Cooperative Extension
https://www.sare.org/Learning-Center/SARE-Project-Products/Western-SARE-Project-Products/Constructing-a-Low-Cost-High-Tunnel
Provides instructions for building a low-cost high tunnel.

Integrated Pest Management (IPM)

Biointensive Integrated Pest Management: Fundamentals of Sustainable Agriculture by Rex Dufour
https://attra.ncat.org/attra-pub/viewhtml.php?id=146
ATTRA also provides a searchable database of biorationals (nonsynthetic and/or less harmful pest-control substances), available at attra.ncat.org/attra-pub/biorationals.

Cornell's Berry Diagnostic Tool, developed by Marvin Pritts
https://blogs.cornell.edu/berrytool/
Got problems with your berries? This site helps you diagnose the culprit for major issues with blueberries, strawberries, and some brambles, and provides links that may lead to a remedy.

New York State Statewide Integrated Pest Management Program (Cornell Cooperative Extension)
www.nysipm.cornell.edu
Along with the University of California site (following), one of the best all-around resources on IPM.

Resource Guide for Organic Insect and Disease Management by Brian Caldwell, et al
https://www.sare.org/Learning-Center/SARE-Project-Products/Northeast-SARE-Project-Products/Resource-Guide-for-Organic-Insect-and-Disease-Management
Excellent guide for use of IPM with organic crops.

Statewide Integrated Pest Management Program (University of California)
www.ipm.ucdavis.edu
Along with the previously listed Cornell site, one of the best all-around resources on IPM; probably my first stop.

Livestock

Livestock are a real investment, so check out primers on your chosen animals under Agriculture Topics, Detailed and (Often) with Further Resources. If you're committed to taking them on, you can find helpful information on several species and breeds on *Hobby Farms* magazine's website (www.hobbyfarms.com), which also includes links to books and magazines for enthusiasts.

Miscellaneous

Cooperative Extension Search
https://search.extension.org/
A site that lets you search for topics across various cooperative extension service publications.

A Global Geospatial Ecosystem Services Estimate of Urban Agriculture
https://agupubs.onlinelibrary.wiley.com/doi/full/10.1002/2017EF000536
Arizona State University/Google-led study by Nicholas Clinton et al in the American Geophysical Union journal *Earth's Future*.

Urban Agriculture: Connecting Producers with Consumers
https://www.emeraldinsight.com/doi/pdfplus/10.1108/BFJ-06-2015-0200
Study of American urban farmer motivations by Carolyn Dimitri, Lydia Oberholtzer, and Andy Pressman in the *British Food Journal*.

New Farmers

Beginning Farmers
https://www.beginningfarmers.org
Offers "information on how to start a farm, planning a new farm, funding resources and finding land to start your farm on." Includes a long list of urban agriculture links and resources.

SeedMoney (formerly Kitchen Gardeners International)
https://seedmoney.org
An organization "offering financial and technical support to a wide variety of public food garden projects."

USDA New Farmers
https://newfarmers.usda.gov
Includes a "discovery" tool to help you whittle down available resources to those most relevant to you.

News, General

City Farmer News
www.cityfarmer.info
The best source for the latest articles and links on urban agriculture.

Organic Farming

The Organic Materials Review Institute
www.omri.org
Provides the latest information about what products are forbidden or permitted in organic production, and under what circumstances.

The Rodale Institute
www.rodaleinstitute.org
Conducts research and provides information on organic agriculture.

Outreach/Research Organizations, Urban Agricultural

Centro Internacional de la Papa/ International Potato Center (CIP)
www.cipotato.org

Concerns for Hunger Organization (ECHO)
www.echonet.org

Consultative Group on International Agricultural Research (CGIAR)
www.cgiar.org

Food for the Cities Program of the Food and Agriculture Organization of the United Nations (FAO)
www.fao.org/fcit/en

The International Development Research Centre (IDRC)
https://www.idrc.caHeifer International
www.heifer.org

Resource Centres on Urban Agriculture and Food Security (RUAF)
www.ruaf.org

UN Habitat
www.unhabitat.org

Plant/Seed Sources

Dave's Garden
http://davesgarden.com
Provides a seed-starting gardener's service as well as user ratings on nurseries and other resources.

Plant Information Online Service of the University of Minnesota Libraries
http://plantinfo.umn.edu
If you want to find sources for a particular plant and know its common or Latin name, you can probably find them through this site.

Plants, Live Sources

Edible Landscaping Online
www.ediblelandscaping.com
Michael McConkey's company, based in zone 7, specializes in "less care" edible plants.

Forest Farm
www.forestfarm.com
Specializes in the Northwest but has an amazing variety of plants.

Plants, Seed Sources

Fedco Seeds
www.fedcoseeds.com
Offers untreated seeds for major vegetable crops.

J.L. Hudson, Seedsman
www.jlhudsonseeds.net
Specializes in rare and ethnobotanically important plants.

Johnny's Selected Seeds
www.johnnyseeds.com
In the author's experience, this is the number-one source of seeds for urban farmers.

Kitazawa Seed Company
www.kitazawaseed.com
Specializes in Asian vegetables.

Seed Savers Exchange
www.seedsavers.org
Features a huge collection of heirloom seeds.

Southern Exposure Seed Exchange
www.southernexposure.com
Specializes in the Mid-Atlantic region.

Territorial Seed Company
www.territorialseed.com
Specializes in the northwestern region.

Plants, Various Information

California Rare Fruit Growers
www.crfg.org
Has information on unusual fruiting plants (especially tropical or subtropical).

Cornell Fruit Resources
http://fruit.cornell.edu
A good place to start for hardier tree fruits, grapes, and berries.

North American Fruit Explorers
www.nafex.org
Another great site for information on unusual fruiting plants.

Plants for a Future Database
www.pfaf.org
Very thorough and includes useful nonfood plants.

Purdue University's Center for New Crops and Plant Products
www.hort.purdue.edu/newcrop
This is probably the first place to look for nontraditional and specialty crop plants.

USDA PLANTS Database
http://plants.usda.gov
Has a huge amount of information on plants, such as geographic spread, noxious status, growth habits, and uses. It's probably the first place to look for nonfood plants.

Soil

Soil and Plant Tissue Testing Laboratory
http://www.umass.edu/soiltest
A good place to have your soil tested is the Soil and Plant Tissue Testing Laboratory of the University of Massachusetts at Amherst. (Be sure to package your soil samples very well so that they do not explode into a dusty mess in postal machinery and cause a scare.)

USDA Guide to Texture by Feel
www.nrcs.usda.gov/wps/portal/nrcs/detail/soils/edu/?cid=nrcs142p2_054311
Offers a flowchart to help you determine soil texture by feel.

USDA List of State Soils
https://www.nrcs.usda.gov/wps/portal/nrcs/detail/soils/edu/?cid=stelprdb1236841
You can find your "official" state soil here.

USDA Web Soil Survey
https://websoilsurvey.sc.egov.usda.gov/App/WebSoilSurvey.aspx
If you're handy with online tools, you can enter your street address and find a whole bunch of information on your local soils (including separates, precipitation, and landforms).

Urban Farmer/Homesteader Blogs

Root Simple
www.rootsimple.com
The blog of The Urban Homestead authors Kelly Coyne and Erik Knutzen.

The Urban Homestead
http://urbanhomestead.org/journal
The blog of the Dervaes family, the original modern urban homesteaders.

Vertical Farming

Prof. Dickson Despommier's Vertical Farm Project
www.verticalfarm.com
"The" site on vertical farming.

Z—Everything Else

The Kansas Center for Sustainable Agriculture and Alternative Crops (KCSAAC)
www.kansassustainableag.org/library.htm
Maintains an excellent list of links to sustainable-agriculture topics organized alphabetically.

INDEX

D

E

F

G

H

I

J

K

L

M

N

T

U

Photo Credits

Illustrations by Tom Kimball unless otherwise noted.

Front cover: Studio Dagdagaz/Shutterstock
Back cover: artur/Shutterstock
Title page: Studio Dagdagaz/Shutterstock

Beth Anderson, InSite Design (for Hilltop Alliance), 378
Charity Burggraaf (www.charitylynne.com), 347
Courtesy Vincent Callebaut Architectures (www.vincent.callebaut.com), 61
Amrita Chatterjee, GREEN Interpreters, www.safeinch.org, 39
Courtesy Canadian Soil Information Service (CanSIS), 381
Courtesy Caroline Clerc, 166
Asahel Curtis, courtesy University of Washington Libraries Asahel Curtis Collection, 341
Courtesy Detroit Mercy and Detroit Free Press, 101
Courtesy EarthBox, 97
Courtesy Edible Landscaping, 234
Courtesy Environmental Protection Agency (EPA), 105
Courtesy Flood/Price/ECHO Staff (www.echonet.org), 69, 141, 143
Thomas J. Fox: 29, 33, 57, 64, 66, 95, 142, 145, 163, 175, 176, 179, 188, 194, 210, 217, 220 (bottom), 229 (bottom), 232, 238, 266 (bottom), 269 (bottom), 272, 276, 281, 282 (bottom), 284, 287, 299, 303, 305, 314, 325, 336, 358
Courtesy Frontera Grill, 300, 301
Kate Arkless Gray (www.flickr.com/radiokate), 45
Courtesy Growing Power, Inc., 117
Courtesy Sherilin Heise, 351
Kristin Lopos, 416
Brooke Salvaggio (www.urbavorefarm.com), 226, 227, 375, 376 (both)
TonyTheTiger, 65
Courtesy United States Department of Agriculture (USDA), 166, 380
VanTucky (CC-BY)/Wikimedia, 351
Courtesy White House, 20
Courtesy Yale Collection of Western America, Beinecke Rare Book and Manuscript Library, 22

The following images courtesy Shutterstock: aastock, 72; Action Sports Photography, 101; Africa Studio, 197; a-image, 80; Valeria Aksakova, 83; allstars, 264; anafotomx, 191; Simone Andress, 52; Anest, 169; AnjelikaGr, 289; Yuri Arcurs, 74; Noah Armonn, 248; artur, 7; aspen rock, 286; AYA images, 114; azure, 223, 252; Darren Baker, 129; Ba_peuceta, 134; Irene Barajas, 23; marilyn barbone, 312; Beata Becla, 280 (top); ALISA BELVILLE, 250; Andi Berger, 247; blueeyes, 216; Bobkeenan Photography, 245; Bork, 323; Sinisa Botas, 135; Alex James Bramwell, 275; Dmitriy Bryndin, 222; Jillian Cain Photography, 31; Tony Campbell, 371; Neil Canon, 181; chillshewa, 71; costall, 139; CP DC Press, 322; Darrell Blake Courtney, 124; Crepesoles, 253; cynoclub, 308; Natalia D, 331; Dallas Events Inc, 122; Chantal de Bruinje, 335; Del Boy, 128; Dionisvera, 364; titov dmitriy, 170; Brian Donaldson, 228; Maria Dryfhout, 280 (bottom); dturphoto, 42; Durden Images, 254; epsilon_lyrae, 219; Joan Ramon Mendo Escoda, 193; Gustavo Miguel Fernandes, 355; Kati Finell, 258; Bill Florence, 283; Flower Studio, 155; forestpath, 333; fotohunter, 292; Tyler Fox, 160; Liv friis-larsen, 246; geertweggen, 369; David Gn, 327; Monika Gniot, 209, 265; Gorilla, 363; Arina P Habich, 24; Miroslav Halama, 361; Alison Hancock, 2-3, 4-5, 106; Hannamariah, 157; Chris Harvey, 8, 237; Rose Hayes, 250; MARGRIT HIRSCH, 109; John Holst, 55; IJPhoto, 132; iwka, 221; Ammit Jack, 358; Tomo Jesenicnik, 259; jgolby, 67; Petr Jilek, 334; Gregory Johnston, 220 (top) ; jstan, 34; jtoddpope, 10; JustASC, 86; Yana Kabangu, 236; Matej Kastelic, 44; David Kay, 116, 249; kd2, 77; sharon kingston, 195; Anne Kitzman, 310; Mr. Klein, 213; Pavel_Klimenko, 239; Joe Klune, 50-51; koosen, 70; Larry Korb, 256, 292; Agatha Koroglu, 320; Vladimir Korostyshevskiy, 89; Slawomir Kruz, 277; kryzhov, 260; Tamara Kulikova, 208; Kuzmaphoto, 288; kzenon, 296; Scott Latham, 332; Karin Hildebrand Lau, 245, 268 (bottom); Varga Levente, 293; Christy Liem, 247; Khoo Si Lin, 199; LianeM, 187; macka, 266 (top); Madlen, 185; Mona Makela, 235, 321; Georgy Markov, 138; mates, 167; Dave McAleavy, 338-339; Olga Miltsova, 131; Tatiana Mirlin, 229 (top); Stefanie Mohr Photography, 243; Elena Moiseeva, 164; MONOPOLY919, 226-227, 272; monticello, 237; Juriah Mosin, 14; motorolka, 367; Naffarts, 370; Maks Narodenko, 262; Nataly Studio, 295; Aksenova Natalya, 352; Olga Nayashkova, 368; NOAA/A. Sapp, Louisiana Universities Marine Consortium; Thomas Nord, 278; Olinchuk, 204; Ollyy, 84-85; Tyler Olson, 27; Marek P, 251; phovoir, 366; picturepartners, 261; Nataliya Peregudova, 263; photogal, 329; pmphoto, 36; Denis Pogostin, 240; Rawpixel.com, 48; Dr Morley Read, 233; Tina Rencelj, 218; rng, 230; Mauro Rodrigues, 270; Zocchi Roberto, 98; Armin Rose, 205; saiko3p, 126-127; sanddebeautheil, 140; sarka, 317; Seasontime, 110; Sally Scott, 328; Svetlana Serebryakova, 214; sevenke, 41, 242, 273; SewCream, 146; Becky Sheridan, 372, 373; Smit, 255; solarseven, 282 (top); Carolina K. Smith MD, 298; akepong srichaichana, 279; taveesak srisomthavil, 302; Alexey Stiop, 269 (top); StockPhotosArt, 268 (top); Laura Stone, 136, 202; Kuttelvaserova Stuchelova, 215; sunsetman, 200; sylv1rob1, 189; szefei, 246; Denis Tabler, 337; Diana Taliun, 345; Tish1, 231; TMsara, 184; Aleksandr Todorovic, 16-17; Toria, 343; Tsekhmister, 350; Al-xVadinska, 324; Liz Van Steenburgh, 291; Videowokart, 285; X-etra, 224; YARUNIV Studio, 225; Yarygin, 304; Larry Ye, 291; Allen W Yoo, 79; yoshi0511, 130; YuRi Photolife, 12; yuris, 290; yurok, 241; claudio zaccherini, 19, 92-93; ZDL, 113; DUSAN ZIDAR, 251; ZoneCreative, 13; ZQFotography, 121; zygomaticus, 244

About the Author

Thomas Fox is the author of two books, *Urban Farming* and *Green Town USA*, which traces the sustainable reinvention of a Kansas town destroyed by a tornado. His nonlinear career path includes stints as a vineyard hand, an "extra" in lineups, an attorney, a professional fact-checker for a major magazine, an independent evaluator, a creative cog in the remorseless gears of educational publishing, and a fundraiser. Through it all, he writes—some fiction, some nonfiction. Mostly short. He lives with his family not far from the world's largest indoor vertical farm. *www.thomasjfox.com*